Reviews of
Environmental Contamination
and Toxicology

VOLUME 213

For further volumes:
http://www.springer.com/series/398

Reviews of Environmental Contamination and Toxicology

Editor
David M. Whitacre

VOLUME 213

 Springer

Coordinating Board of Editors

ISSN 0179-5953
ISBN 978-1-4614-2873-2   ISBN 978-1-4419-9860-6 (eBook)
DOI 10.1007/978-1-4419-9860-6
Springer New York Dordrecht Heidelberg London

Springer is part of Springer Science+Business Media (www.springer.com)

Foreword

International concern in scientific, industrial, and governmental communities over traces of xenobiotics in foods and in both abiotic and biotic environments has justified the present triumvirate of specialized publications in this field: comprehensive reviews, rapidly published research papers and progress reports, and archival documentations. These three international publications are integrated and scheduled to provide the coherency essential for nonduplicative and current progress in a field as dynamic and complex as environmental contamination and toxicology. This series is reserved exclusively for the diversified literature on "toxic" chemicals in our food, our feeds, our homes, recreational and working surroundings, our domestic animals, our wildlife, and ourselves. Tremendous efforts worldwide have been mobilized to evaluate the nature, presence, magnitude, fate, and toxicology of the chemicals loosed upon the Earth. Among the sequelae of this broad new emphasis is an undeniable need for an articulated set of authoritative publications, where one can find the latest important world literature produced by these emerging areas of science together with documentation of pertinent ancillary legislation.

Research directors and legislative or administrative advisers do not have the time to scan the escalating number of technical publications that may contain articles important to current responsibility. Rather, these individuals need the background provided by detailed reviews and the assurance that the latest information is made available to them, all with minimal literature searching. Similarly, the scientist assigned or attracted to a new problem is required to glean all literature pertinent to the task, to publish new developments or important new experimental details quickly, to inform others of findings that might alter their own efforts, and eventually to publish all his/her supporting data and conclusions for archival purposes.

In the fields of environmental contamination and toxicology, the sum of these concerns and responsibilities is decisively addressed by the uniform, encompassing, and timely publication format of the Springer triumvirate:

Reviews of Environmental Contamination and Toxicology [Vol. 1 through 97 (1962–1986) as Residue Reviews] for detailed review of articles concerned with any aspects of chemical contaminants, including pesticides, in the total environment with toxicological considerations and consequences.

Bulletin of Environmental Contamination and Toxicology (Vol. 1 in 1966) for
rapid publication of short reports of significant advances and discoveries
in the fields of air, soil, water, and food contamination and pollution as
well as methodology and other disciplines concerned with the introduction,
presence, and effects of toxicants in the total environment.

Archives of Environmental Contamination and Toxicology (Vol. 1 in 1973)
for important complete articles emphasizing and describing original experimental or theoretical research work pertaining to the scientific aspects of
chemical contaminants in the environment.

Manuscripts for *Reviews* and the *Archives* are in identical formats and are peer
reviewed by scientists in the field for adequacy and value; manuscripts for the
Bulletin are also reviewed but are published by photo-offset from camera-ready copy
to provide the latest results with minimum delay. The individual editors of these
three publications comprise the joint Coordinating Board of Editors with referral
within the board of manuscripts submitted to one publication but deemed by major
emphasis or length more suitable for one of the others.

Coordinating Board of Editors

Preface

The role of *Reviews* is to publish detailed scientific review articles on all aspects of environmental contamination and associated toxicological consequences. Such articles facilitate the often complex task of accessing and interpreting cogent scientific data within the confines of one or more closely related research fields.

In the nearly 50 years since *Reviews of Environmental Contamination and Toxicology* (formerly *Residue Reviews*) was first published, the number, scope, and complexity of environmental pollution incidents have grown unabated. During this entire period, the emphasis has been on publishing articles that address the presence and toxicity of environmental contaminants. New research is published each year on a myriad of environmental pollution issues faced by people worldwide. This fact, and the routine discovery and reporting of new environmental contamination cases, creates an increasingly important function for *Reviews*.

The staggering volume of scientific literature demands remedy by which data can be synthesized and made available to readers in an abridged form. *Reviews* addresses this need and provides detailed reviews worldwide to key scientists and science or policy administrators, whether employed by government, universities, or the private sector.

There is a panoply of environmental issues and concerns on which many scientists have focused their research in past years. The scope of this list is quite broad, encompassing environmental events globally that affect marine and terrestrial ecosystems; biotic and abiotic environments; impacts on plants, humans, and wildlife; and pollutants, both chemical and radioactive; as well as the ravages of environmental disease in virtually all environmental media (soil, water, air). New or enhanced safety and environmental concerns have emerged in the last decade to be added to incidents covered by the media, studied by scientists, and addressed by governmental and private institutions. Among these are events so striking that they are creating a paradigm shift. Two in particular are at the center of ever-increasing media as well as scientific attention: bioterrorism and global warming. Unfortunately, these very worrisome issues are now superimposed on the already extensive list of ongoing environmental challenges.

The ultimate role of publishing scientific research is to enhance understanding of the environment in ways that allow the public to be better informed. The term "informed public" as used by Thomas Jefferson in the age of enlightenment conveyed the thought of soundness and good judgment. In the modern sense, being "well informed" has the narrower meaning of having access to sufficient information. Because the public still gets most of its information on science and technology from TV news and reports, the role for scientists as interpreters and brokers of scientific information to the public will grow rather than diminish. Environmentalism is the newest global political force, resulting in the emergence of multinational consortia to control pollution and the evolution of the environmental ethic. Will the new politics of the twenty-first century involve a consortium of technologists and environmentalists, or a progressive confrontation? These matters are of genuine concern to governmental agencies and legislative bodies around the world.

For those who make the decisions about how our planet is managed, there is an ongoing need for continual surveillance and intelligent controls to avoid endangering the environment, public health, and wildlife. Ensuring safety-in-use of the many chemicals involved in our highly industrialized culture is a dynamic challenge, for the old, established materials are continually being displaced by newly developed molecules more acceptable to federal and state regulatory agencies, public health officials, and environmentalists.

Reviews publishes synoptic articles designed to treat the presence, fate, and, if possible, the safety of xenobiotics in any segment of the environment. These reviews can be either general or specific but properly lie in the domains of analytical chemistry and its methodology, biochemistry, human and animal medicine, legislation, pharmacology, physiology, toxicology, and regulation. Certain affairs in food technology concerned specifically with pesticide and other food-additive problems may also be appropriate.

Because manuscripts are published in the order in which they are received in final form, it may seem that some important aspects have been neglected at times. However, these apparent omissions are recognized, and pertinent manuscripts are likely in preparation or planned. The field is so very large and the interests in it are so varied that the editor and the editorial board earnestly solicit authors and suggestions of underrepresented topics to make this international book series yet more useful and worthwhile.

Justification for the preparation of any review for this book series is that it deals with some aspect of the many real problems arising from the presence of foreign chemicals in our surroundings. Thus, manuscripts may encompass case studies from any country. Food additives, including pesticides, or their metabolites that may persist into human food and animal feeds are within this scope. Additionally, chemical contamination in any manner of air, water, soil, or plant or animal life is within these objectives and their purview.

Manuscripts are often contributed by invitation. However, nominations for new topics or topics in areas that are rapidly advancing are welcome. Preliminary communication with the editor is recommended before volunteered review manuscripts are submitted.

Summerfield, North Carolina   David M. Whitacre

Contents

Contributors

Folahan A. Adekola Department of Chemistry, Faculty of Science, University of Ilorin, P.M.B.1515 Ilorin, Nigeria, faadekola@yahoo.fr

Jean-Claude Amiard Service d'Ecotoxicologie – "Mer, Molécules, Santé", EA 2160, Université de Nantes, Nantes, France, jean-claude.amiard@univ-nantes.fr

Nathalie Arnich Direction Santé Alimentation, Agence nationale de sécurité sanitaire de l'alimentation, de l'environnement et du travail (ANSES), 94701 Maisons-Alfort, France, nathalie.arnich@anses.fr

Pierre-Marie Badot UMR Chrono-environnement, CNRS/Université de Franche-Comté usc INRA, F-25030 Besançon cedex, France, pierre-marie.badot@univ-fcomte.fr

Didier Claisse Département Biogéochimie et Ecotoxicologie, ROCCH, IFREMER, BP 21105 – 44 311 Nantes Cedex 3, France, didier.claisse@ifremer.fr

Camille Dumat EcoLab (Laboratoire d'écologie fonctionnelle), INP-ENSAT, 31326, Castanet-Tolosan, France; EcoLab (Laboratoire d'écologie fonctionnelle), UMR 5245 CNRS-INP-UPS, 31326 Castanet-Tolosan, France, camille.dumat@ensat.fr

Olalekan S. Fatoki Department of Chemistry, Faculty of Applied Sciences, Cape Peninsula University of Technology, Cape Town 7535, South Africa, fatokio@cput.ac.za

Marielle Guéguen Unité des microorganismes d'intérêt laitier et alimentaire EA 3213, UFR ICORE 146, Université de Caen-Basse Normandie, 14032 Caen Cedex 5, France, marielle.gueguen@unicaen.fr

Thierry Guérin Unité Contaminants inorganiques et minéraux de l'environnement (CIME), Agence nationale de sécurité sanitaire de l'alimentation, de l'environnement et du travail (ANSES), Laboratoire de sécurité des aliments, ANSES – LSA, 94706 Maisons-Alfort Cedex, France, thierry.guerin@anses.fr

Florian Keil Institute for Social-Ecological Research ISOE, 60486 Frankfurt am Main, Germany, florian.keil@alice-dsl.net

Lina Kiaune Department of Pesticide Regulation, California Environmental Protection Agency, Sacramento, CA 95812-4015, USA, lkiaune@cdpr.ca.gov

Jörg Oehlmann Department of Aquatic Ecotoxicology, Institute of Ecology, Evolution and Diversity, Goethe University Frankfurt am Main, 60323 Frankfurt am Main, Germany, oehlmann@bio.uni-frankfurt.de

Hussein K. Okoro Department of Chemistry, Faculty of Applied Sciences, Cape Peninsula University of Technology, Cape Town 7535, South Africa, okoroowo@yahoo.com; okorohk@cput.ac.za

Beatrice Opeolu Department of Chemistry, Faculty of Applied Sciences, Cape Peninsula University of Technology, Cape Town 7535, South Africa, opeolubt@cput.ac.za

Eric Pinelli EcoLab (Laboratoire d'écologie fonctionnelle), INP-ENSAT, 31326 Castanet-Tolosan, France; EcoLab (Laboratoire d'écologie fonctionnelle), UMR 5245 CNRS-INP-UPS, 31326 Castanet-Tolosan, France, pinelli@ensat.fr

Bertrand Pourrut LGCgE, Equipe Sols et environnement, ISA, 59046 Lille Cedex, France; Université de Toulouse, Toulouse, France, b.pourrut@isa-lille.fr

Ulrike Schulte-Oehlmann Department of Aquatic Ecotoxicology, Institute of Ecology, Evolution and Diversity, Goethe University Frankfurt am Main, 60323 Frankfurt am Main, Germany, schulte-oehlmann@bio.uni-frankfurt.de

Muhammad Shahid EcoLab (Laboratoire d'écologie fonctionnelle), INP-ENSAT, 31326 Castanet-Tolosan, France; EcoLab (Laboratoire d'écologie fonctionnelle), UMR 5245 CNRS-INP-UPS, 31326 Castanet-Tolosan, France, shahidzeeshan@gmail.com

Nan Singhasemanon Department of Pesticide Regulation, California Environmental Protection Agency, Sacramento, CA 95812-4015, USA, nsinghasemanon@cdpr.ca.gov

Reinette G. Snyman Department of Biodiversity and Conservation, Faculty of Applied Sciences, Cape Peninsula University of Technology, Cape Town 8000, South Africa, snymanr@cput.ac.za

Jean-Paul Vernoux Unité des microorganismes d'intérêt laitier et alimentaire EA 3213, UFR ICORE 146, Université de Caen-Basse Normandie, 14032 Caen Cedex 5, France, jean-paul.vernoux@unicaen.fr

Peter Winterton Université Paul Sabatier, 31062 Toulouse, France, peter.winterton@univ-tlse3.fr

Bhekumusa J. Ximba Department of Chemistry, Faculty of Applied Sciences, Cape Peninsula University of Technology, Cape Town 7535, South Africa, ximbab@cput.ac.za

Pesticidal Copper (I) Oxide: Environmental Fate and Aquatic Toxicity

Lina Kiaune and Nan Singhasemanon

Contents

1 Introduction

Copper oxide is used in agriculture as a fungicide to protect coffee, cocoa, tea, banana, citrus, and other plants from major fungal leaf and fruit diseases such as blight, downy mildew, and rust (HSDB 2008). Copper oxide is used as an active ingredient in various pesticidal formulations. After the ban of tributyltin (TBT), in the late 1980s, the use of copper oxide in antifouling paint products increased. These products protect boat and ship hulls against biofouling by marine organisms. There are currently 209 pesticide products registered in California that use copper

N. Singhasemanon (✉)
Department of Pesticide Regulation, California Environmental Protection Agency,
Sacramento, CA 95812-4015, USA
e-mail: nsinghasemanon@cdpr.ca.gov

D.M. Whitacre (ed.), *Reviews of Environmental Contamination and Toxicology*,
Reviews of Environmental Contamination and Toxicology 213,
DOI 10.1007/978-1-4419-9860-6_1, © Springer Science+Business Media, LLC 2011

oxide as an active ingredient (CDPR 2009a). Examples of registered pesticide products include the following: 3M Copper Granules, Americoat 275E Antifouling Red, Copper Shield 45, Nordox, Super KL K90 Red, Ultra 3559 Green, and others.

Although copper is an effective biocide, it may also affect non-target organisms and pose environmental concerns. Copper may be washed into the aquatic environment from agricultural and urban application sites and may enter water when used as a biocide in antifouling paint formulations. The latter use may constitute a major copper pollution contributor to California marinas, because antifouling paints continually leach from and are regularly scrubbed off boat hulls, thus releasing copper-containing paint residues into the surrounding water and sediment. The resulting copper concentrations may potentially be high enough to threaten aquatic organisms.

Copper (Cu) is a naturally occurring element. Its average abundance in the earth's crust is about 50 parts per million (ppm) (U.S. DHHS 2004). Copper is a transitional metal and occurs in nature in four oxidation states: elemental copper Cu (0) (solid metal), Cu (I) cuprous ion, Cu (II) cupric ion, and rarely Cu (III) (Georgopoulos et al. 2001).

Copper is also a trace element that is needed for proper functioning of many enzymes in biological systems. At least 21 copper-containing enzymes are known, all of which function as redox catalysts (e.g., cytochrome oxidase, monoamine oxidase) or dioxygen carriers (e.g., hemocyanin) (Weser et al. 1979). Excess copper concentrations, on the other hand, retard organisms' vital processes by inactivating enzymes and by precipitating cytoplasmic proteins into metallic proteinates (Long 2006). Exposure to copper-containing compounds precedes the modern era; such compounds have been used as pesticides for centuries and are still being used today in various insecticide, fungicide, herbicide, algaecide, and molluscicide formulations.

In this chapter, we review the environmental fate and effects of copper oxide, with special attention provided to surface waters: freshwater, saltwater, and brackish water. Since copper is a natural element, its speciation, environmental fate, and toxicity are complex and differ from that of organic pesticides. In water, Cu (II) (or Cu^{2+}) is the most prevalent form of copper (Georgopoulos et al. 2001). Therefore, in this review we will primarily focus on this ionic species. Additionally, because of rapid advances in nanotechnology and potential developments of nanopesticides, we will also address the current state of knowledge on the environmental fate and toxicity of nanocopper.

1.1 Molecular Structure

Copper (I) oxide is a mineral that has cubic structure. In the lattice structure, copper has two neighboring oxygen atoms, and oxygen has four neighboring copper atoms (Web Elements 2009). Copper (I) oxide is an IUPAC name; however, in this chapter synonymous names like copper oxide and cuprous oxide will be used interchangeably.

1.2 Physical and Chemical Properties

Copper oxide dissolves in strong acids, ammonium hydroxide, and aqueous ammonia and its salts (Goh 1987). Copper oxide is insoluble in water, organic solvents, and dilute acid unless an oxidizing agent is present (U.S. DHHS 2004) (Table 1).

1.3 Use in California

The majority of copper oxide is used in agriculture on nuts, citrus, apples, lettuce, olives, berries, spices, and other commodities (Fig. 1) (CDPR 2009b). The pounds of copper oxide used in 2007 are similar to those used in 1997, and amounts used have generally decreased since 2005. The amounts of copper (I) oxide used between 1993 and 1995 were unavailable and therefore represent data gaps in the chart depicted in Fig. 1.

Copper oxide is the most popular biocide used in antifouling paints today, appearing in about 90% of products registered in California (Singhasemanon et al. 2009). Antifouling uses of copper oxide include commercial and non-commercial applications for boat and ship hulls and miscellaneous applications such as underwater structures and piers. Efforts to estimate the total use in pounds of copper oxide for these purposes are challenging, because the use of copper as an antifouling agent is not required to be reported.

Table 1 Physical and chemical properties of copper (I) oxide

IUPAC name	Copper (I) oxide
Synonyms	Copper oxide; cuprous oxide; dicopper oxide
CAS number	1317-39-1
Molecular formula	Cu_2O
Molecular weight (g/mol)	143.09
Appearance	Yellow, red, or brown crystalline powder
Odor	None
Boiling point STP[a] (°C)	1800
Melting point STP (°C)	1235
Density (g/cm^3)	6
Refractive index	2.705
Vapor pressure	Negligible
Solubility	Soluble in dilute mineral acid to form copper (I) salt or copper (II) salt plus metallic copper; aqueous ammonia and its salts
	Insoluble in water and organic solvents
Stability	Stable in dry air; in moist air oxidizes to cupric oxide
Kow	Not applicable

Sources: CDPR (1991); Goh (1987); HSDB (2008); ILO (2008) (ICSC: 0421)
[a]STP: standard temperature and pressure

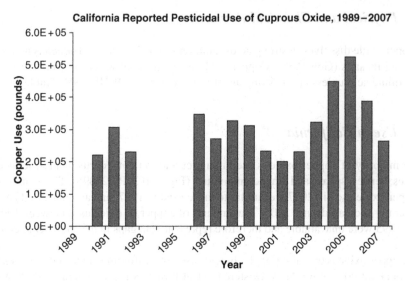

Fig. 1 Cuprous oxide total annual use in California between 1989 and 2007. *Source*: CDPR (2009b)

2 Environmental Fate

2.1 Copper Speciation in Surface Waters

In the water column, copper displays a complex biogeochemical and speciation cycle (Fig. 2). Important factors that determine copper's environmental fate relate to interactions between the metal and the physical/chemical properties of the water column. Seawater, water of increasing salinity in estuaries, and freshwater in rivers and lakes may have different copper speciation outcomes. These outcomes influence metal bioavailability and, thus, the toxicity to the aquatic organisms.

Copper oxide dissociates in water, and the most prevalent copper oxidation state is Cu^{2+} (cupric ion) (Georgopoulos et al. 2001). Cu^{2+} is also a form primarily responsible for coppers' biocidal effects. Thus, the following discussion will largely refer to water solubilized copper.

Copper can exist adsorbed to dissolved molecules or to particulate matter and is referred to collectively as the total copper (TCu) pool (Fig. 3). Even though copper adsorbs to particulate matter, it interacts most strongly with dissolved components in the water (Muller 1996). Hence, in speciation studies, total dissolved copper (TDCu) concentrations, sometimes referred to as dissolved copper (DCu), is the entity that is conventionally measured. TDCu is functionally determined by the filter pore size. Copper passing through a 0.45 μm or smaller filter pore size is considered to be dissolved.

TDCu can be further separated into labile copper (LCu) and organically complexed fractions. In this chapter, the term LCu means bioavailable copper

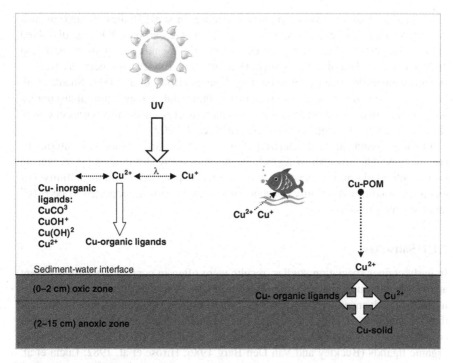

Fig. 2 Aquatic fate diagram for copper oxide (POM is particulate organic matter)

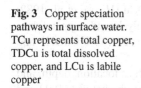

Fig. 3 Copper speciation pathways in surface water. TCu represents total copper, TDCu is total dissolved copper, and LCu is labile copper

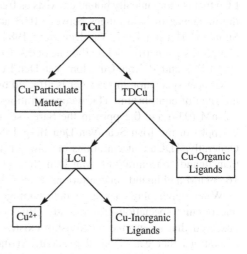

and includes both free hydrated copper ions (Cu^{2+}) and inorganically complexed species. Organically complexed copper, on the other hand, is considered inert or non-bioavailable to biological organisms.

Organically complexed copper is bound to organic ligands. "Organic ligands" is a generic term describing heterogeneous molecules that are ubiquitous in water.

Their binding may or may not be metal specific. In some studies, organic ligands are interchangeably referred to as dissolved organic carbon (DOC) or dissolved organic matter (DOM). The sources of organic ligands are both natural and anthropogenic. Natural sources include humic and fulvic substances, as well as microorganism-produced copper-binding ligands (Buck et al. 2007; Shank et al. 2004). Anthropogenic sources include urban, industrial, and agricultural discharges and runoffs that carry organic molecules such as ethylenediaminetetraacetic acid (EDTA) and nitrilotriacetic acid (NTA) (Buck et al. 2007).

Organic ligands are characterized as weak, strong, and, sometimes, intermediate strength. The strength of ligands is experimentally determined by measuring their conditional stability constants, which reflect copper-ligand binding affinity. For example, conditional stability constants for weak ligands are reported to be $\sim 10^9$ and for strong ligands $\sim 10^{13}$.

2.1.1 Saltwater

In early copper speciation studies, results were often in conflict, because different analysis techniques and detection capabilities of analytical instrumentation were used. With advanced instrumentation, a more uniform picture emerged, showing that the majority of TDCu in seawater is organically complexed. Most authors of copper speciation studies have concluded that about 89–99% of TDCu is bound to organic ligands (Buckley and Van Den Berg 1986; Hirose et al. 1982; Lucia et al. 1994; Suda and Hanson 1987). Therefore, only a small fraction of TDCu constitutes LCu that is inorganically bound or exists as free Cu^{2+} ions. Among different studies, the incidence of LCu ranges between 0.03 and 6% of TDCu (Hirose et al. 1982; Suda and Hanson 1987; Van Den Berg 1984). Inorganically bound copper forms complexes primarily with carbonate (60% $CuCO_3$), hydroxide (32% $CuOH^+$ and $Cu(OH)_2$), and 4% free Cu^{2+} ion (Van Den Berg 1984).

Copper speciation in seawater is affected by factors such as location, depth, and the state of equilibrium. TDCu concentrations were found to be between 0.92 and 3.2 nM (0.06 and 0.2 ppb) in the Sargasso and North seas and 16–39 nM (1.0–2.5 ppb) in the Irish Sea (Van Den Berg 1984; Van Den Berg and Donat 1992). Generally, TDCu concentrations in seawater have been observed to be much lower than ligand concentrations. Van Den Berg (1984) found TDCu concentrations of 16–39 nM and ligand concentrations of 58–156 nM.

Water depth plays a role in determining metal and ligand concentrations. In one organic speciation profile study (Buckley and Van Den Berg 1986), it was observed that an inverse relationship existed between free Cu^{2+} and ligand concentrations that was depth dependent. At the surface, free Cu^{2+} concentrations were low (2×10^{-13} M) and ligand concentrations were high (maximum 1800 nM). With increased depth, ligand concentrations dropped to 6–20 nM and free Cu^{2+} concentration increased to 7×10^{-13} M.

In the majority of copper speciation studies, it is assumed that free copper in the water is in a state of equilibrium. The state of equilibrium predicts that free copper in water exists primarily in the Cu^{2+} oxidation state. Moffett and Zilka (1983), however, challenged this assumption. According to these authors, biochemical, photochemical, or thermodynamic processes create a non-equilibrium environment, in which copper redox chemistry may become an important part of copper speciation. For example, in the photic zone of the ocean, sunlight-generated free radicals like superoxide and hydrogen peroxide can reduce Cu^{2+} to Cu^+. Thus, the thermodynamic model used by the authors predicted that 20–50% of total copper would be Cu^+. Other factors that affect reduction to Cu^+ are pH, dissolved oxygen concentration, ligands, and reducing species (Moffett and Zilka 1983).

2.1.2 Brackish Water

The salinity of brackish water is less than that of seawater. Brackish water most commonly occurs in estuaries, where fresh river water meets seawater. Complex copper speciation takes place in this mixing zone.

Upon reaching water of increasing salinity, copper is largely sequestered by forming complexes with organic ligands (Apte et al. 1990; Buck and Bruland 2005; Hurst and Bruland 2005; Kozelka and Bruland 1998; Muller 1996). Apte et al. (1990) found TDCu concentrations to be higher at the freshwater end. The TDCu concentrations displayed a linear decrease with an increase in water salinity, changing from 76.4 nM in the freshwater samples to 6.8 nM in saline samples.

Seasonal and temperature fluctuations also affect TDCu and LCu levels. Beck and Sanudo-Wilhelmy (2007) studied seasonal TDCu cycling in the Long Island Sound, New York. They observed that TDCu levels did not vary greatly between spring and summer. Surface LCu concentrations, on the other hand, showed seasonal cycling to be higher in the summer. In another study, Jones and Bolam (2007) observed increased TDCu levels from winter to late summer and decreased levels during the autumn and winter, in UK marinas. Unlike Beck and Sanudo-Wilhelmy (2007), this group observed that the LCu fraction remained fairly constant throughout the year.

Although Jones and Bolam (2007) indicated that the natural environment has sufficient buffering capacity to keep LCu concentrations low, Beck and Sanudo-Wilhelmy (2007) linked the LCu fluctuation to water temperature and dissolved oxygen levels. At water temperatures above 21°C, LCu concentrations in the bottom waters increased exponentially. This indicates that copper remobilization was occurring and could explain the increased surface LCu levels during the summer. High levels of LCu were also observed under low oxygen conditions. The authors believe that there is a potential for copper remobilization if water temperatures rise and dissolved oxygen concentrations decrease as a result of global warming.

As occurred in seawater, the conclusion reached from the majority of the brackish water copper speciation studies was that 97–99.99% of TDCu in estuaries is organically complexed to ligands (Apte et al. 1990; Buck and Bruland 2005; Hurst and Bruland 2005; Kozelka and Bruland 1998; Muller 1996). The conditional stability constants of the ligands in brackish water are comparable to those of seawater. Organic ligand concentrations generally exceed TDCu concentrations, as well (Buck and Bruland 2005; Hurst and Bruland 2005; Kozelka and Bruland

1998). For example, TDCu levels in San Francisco Bay ranged between 17.9 and 49.6 nM and strong ligand concentrations were between 22 and 265 nM (Buck and Bruland 2005). As a result, free Cu^{2+} concentrations were low.

Since organic ligand levels consistently exceed TDCu concentrations, only a small percentage of copper constitutes the LCu fraction. Hence, there is a risk of overestimating the levels of copper that are available to cause toxicity. Buck and Bruland (2005) derived a saturation curve-shaped relationship between TDCu and Cu^{2+}. They estimated that for free copper to reach toxic levels of 10^{-11} M, the TDCu must be at least 100 nM. A similar finding that TDCu measurements tend to overestimate copper toxicity was also noted by Jones and Bolam (2007). Their calculated LCu/TDCu ratio predicted that TDCu overestimates the toxicity risk by a factor of 4.

2.1.3 Freshwater

Among saltwater, brackish water, or freshwater types, copper speciation in the latter appears to be the least studied and generates the most debated results. Water characteristics in rivers and lakes differ in ways that depend largely on the geochemistry of their particular location. Therefore, water properties (e.g., pH, hardness, and alkalinity) affect speciation and may produce different results among studies performed at different locations. The use of different techniques and inconsistent terminology can also produce different measurements or interpretations of results.

Copper speciation in freshwater is predominantly controlled by the TDCu that binds to organic ligands (Hoffman et al. 2007). According to the authors, almost all dissolved copper (>99.99%) is bound to strong ligands in river water, which produces free Cu^{2+} concentrations that are in the picomolar range ($0.9–6.5 \times 10^{-15}$ M). Most authors agree that, in freshwater, ligand concentrations consistently exceed those of TDCu.

The conditional stability constants of the freshwater ligands differ from those found in saltwater and estuaries. Hoffman et al. (2007) reported ligand conditional stability constants above 10^{13}. Additionally, the ligands had a higher affinity for copper than for other metals. This may indicate the existence of copper-specific ligands in freshwater systems. In contrast, Wang and Chakrabarti (2008) and Pesavento et al. (2003) reported the conditional stability constants for very strong ligands to be $\sim 10^{20}$ and 10^{17}, respectively. These numbers are many folds higher than the estimated strong ligand conditional stability constants found in saltwater and estuaries ($\sim 10^{13}$). Because of such inconsistent results, more studies in freshwater are needed to determine the nature and origin of such ligands.

In freshwater, as in saltwater and brackish water, organic complexation generally dominates copper speciation and, thus, toxicity. However, in freshwater, parameters like pH, alkalinity, and hardness have significant effects on copper speciation, as well. Gundersen and Steinnes (2003) studied eight rivers in Norway and determined that pH had the most significant influence on metal speciation. At low pH levels, most of the copper was dissolved, and at high pH levels, Cu occurred predominantly in colloidal or particulate form. Consequently, in river water of pH

3.1, almost all copper was in the dissolved fraction, and at a pH range of 6.9–7.2 (pH neutral rivers), all three fractions (dissolved, colloidal, and particulate) occurred in significant amounts. From these results, it can be inferred that LCu concentrations may be higher in acidic water.

Alkalinity is related to the capacity of water to neutralize strong acids (Snoeyink and Jenkins 1980a). Alkalinity (alkalinity ions HCO^{3-} and CO_3^{2-}) and water hardness (Ca^{2+} and Mg^{2+} ion concentrations) are related and usually increase or decrease together. According to Snoeyink and Jenkins (1980b), in a carbonate-buffered water system with pH below 6.5, the predominant copper species is Cu^{2+}. In the pH range of 6.5–9.5 (the pH range of most waters), $CuCO_3$ is the predominant copper species. Hence, in most waters, copper forms copper carbonate complexes. Moreover, at pH 7 an increase in alkalinity from 50 to 250 mg/L (as $CaCO_3$) decreases the Cu^{2+} levels from 25 to 9% of the total copper present. These results indicate that copper is more bioavailable and more toxic in soft, less alkaline water than in hard, more alkaline water.

2.2 Copper Speciation in Sediment

Although sediments tend to accumulate heavy metals, mass-balance estimates suggest that their remobilization could be a major source of some toxic metals in the water column (Beck and Sanudo-Wilhelmy 2007). Therefore, it is important to understand copper speciation in sediment pore water. Sediment pore water fills the spaces between grains of sediment. At the sediment–water boundary, physical, chemical, and biological changes take place (Fig. 2). Processes (physical, chemical, and biological) that bring about changes in the sediment (following its deposition) are referred to as diagenesis (Berner 1980). Copper speciation in pore water is influenced by diagenetic processes and depends on factors such as oxygen levels, temperature, and sediment type. Before addressing the specifics of copper speciation in sediment, it is important to explain how copper cycles globally to end up at the bottom of the water column.

The geochemical cycling of copper in the water column is linked to the cycling of organic carbon (Widerlund 1996). In surface water, copper adsorbs to scavenging particulate organic matter (POM) and, with the downward flux, eventually settles out. The settling action results in the formation of a thin, carbon-rich layer at the sediment–water interface (Klinkhammer et al. 1982). How much TCu settles along with POM depends on the location and season. Chester et al. (1988) reported that detritus-associated Cu comprised about 60% of TCu along the Atlantic coast, but only 18% in the open ocean. These differences reflect differences in oceanic biological activity. Landing and Feely (1981) observed increased copper flux during the summer algae bloom in the Icy Bay, Gulf of Alaska. Similarly, Helland and Bakke (2002) reported higher Cu–POM fluxes near the river mouth during a spring flood in the Gloma Estuary, Norway. Because many factors interplay in the global carbon cycling, the study authors reported that 10–50% of TCu is bound to suspended POM and settle out (Chester et al. 1988; Helland and Bakke 2002).

Upon settling, diagenetic processes break the Cu–POM association and copper is liberated into the sediment pore water; herein, copper may partition back into the water column or into the solid sediment phase. The residence time for copper in the pore water is approximately 2.1–10 days (Widerlund 1996).

The availability of oxygen determines whether copper is recycled back into the water column or is removed by precipitation. Experimental results from Widerlund (1996) suggest that oxic conditions (0–2 cm depth) play a major role in copper recycling, whereas anoxic (2–15 cm depth) conditions result in copper removal (Fig. 2). Other study results have also shown that, during early diagenesis, aerobic decomposition controls copper release back to the water column (Gerringa 1990; Skrabal et al. 2000). Gerriga et al. (1990) reported two kinds of aerobic degradation: fast and slow. Fast degradation was characterized by the rapid decline in POM concentration and high oxygen consumption. During fast degradation, copper in the sediment, which was relatively strongly bound, became relatively weakly bound. Slow degradation of POM was reflected by the transformation of ammonia into nitrite and nitrite into nitrate. It is through such transformations that copper, derived from degrading organic constituents of sediment, is continuously released into the sediment pore water. Copper concentrations were reported to be 10 times higher in the top 2 cm of the sediment pore water than in the overlying bottom waters (Klinkhammer et al. 1980). At the deeper levels of sediment pore water, where anoxic conditions prevail, copper is captured into the solid-phase sediment by precipitating metal sulfides (Skrabal et al. 2000; Widerlund 1996).

Temperature plays an important role in copper cycling. Seasonal temperature fluctuations can affect sediment conditions by changing biological activity and oxygen levels, thus influencing copper benthic flux. Widerlund (1996) noted that removal of dissolved copper from the pore water into the solid-phase sediment was temperature dependent. In September (core temperature 8°C), copper flux into the sediment was twice as high as in April (1°C). Hence, at higher water temperatures more copper is removed from the sediment pore water by precipitation. The authors also noted that since decomposition of the organic matter is temperature dependent, cold water promotes higher sediment accumulation and results in a more rapid burial of reactive (non-decomposed) organic matter.

In addition to oxygen availability and water temperature, sediment type and micro-flora and microfauna may play a role in the speciation of copper in sediment pore water. Goh and Chou (1997) observed that finer sediment has higher surface area onto which copper can adsorb. Skrabal et al. (2000) investigated the distribution of TDCu at two distinct locations: one with sulfidic muddy sediment, dominated by seasonal anoxia and poor biodiversity, and the other with sandy silt, dominated by extensive bioturbation and richness in benthic organisms. The authors determined that copper precipitates out as metal sulfides in anoxic, muddy sediment conditions. The study also allowed the authors to predict that, under these conditions, copper exists predominantly in the Cu^+ oxidation state and is bound to sulfur-containing organic and inorganic ligands. In contrast, in oxic and biodiverse sandy sediment, the Cu^{2+} oxidation state dominates TDCu speciation. Under oxic conditions, the release of copper is controlled by the aerobic decomposition rate of organic matter.

When copper disassociates from POM and is liberated into the sediment pore water, it enters the TDCu phase. As occurs in the water column, TDCu speciation and bioavailability in the sediment pore water are controlled by strong and weak organic ligands (Gerriga et al. 1991; Skrabal et al. 2000). Since sediments are rich in organic matter, they provide a large pool of ligands to the pore water that is available for metal complexation. In addition, Skrabal et al. (2000) suggest that sediment pore water also supplies as much as 10–50% of copper complexing ligands to the overlying water column. Skrabal et al. (2000) found ligands to always be in large excess relative to TDCu concentrations in the sediment pore water. As a result, 87–99.9% of copper exists as organic complexes, and free inorganic copper concentrations are low.

Organic ligands in the sediment pore water, in contrast to their water column analogs, are subject to much faster biological degradation. Gerriga et al. (1991) reported differences in ligand degradation rates. Strong ligands were subject to oxidation, and their concentrations decreased faster than did weak ligand concentrations. Weak ligands were more resistant to degradation. After strong ligand concentrations became depleted, weaker ligands began to dominate TDCu speciation. This resulted in a sharp increase of free copper concentration (from 10^{-12} to 10^{-9} M) (Gerriga et al. 1991). Although sediments are a rich source of organic ligands, when TDCu speciation is dominated by weak ligands, copper may be much more bioavailable and, thus, more toxic to the aquatic organisms.

2.3 Copper Speciation in Soil

Copper oxide is used in agriculture to protect various crops from fungal diseases. Soil is also a major repository of copper. Thus, copper pollution can affect soil-dwelling organisms and plants and make its way to the food chain. Understanding copper speciation in soil is important to a better understanding of its effects on the soil ecosystem. The factors that control the environmental fate of copper in soil include the organic and inorganic content and the pH.

Copper persists in the topsoil and generally accumulates in the upper 15 cm (Rodriguez-Rubio et al. 2003). Today, elevated copper concentrations are found in the vicinity of former smelters and chemical spill areas that did occur, or may have occurred, decades previously (Kabala and Singh 2001; McBride and Bouldin 1984). The results from several studies show that copper preferentially associates with soil organic matter (Boudesocque et al. 2007; Jacobson et al. 2007; Liu and Wang 2004; Strawn and Baker 2008). In the Liu and Wang (2004) speciation study, 50% of copper in the contaminated soil was associated with organic matter, 28% formed $CuCO_3$, 11% Cu_2O, and 11% CuO. Boudescque et al. (2007) determined that the copper associated with soil organic matter is formed via inner sphere complexes, which occur when copper ions adsorb directly to the organic particle in the soil. Because of the strength of such complexes, organic matter plays an important role in determining the degree of mobility and bioavailability of copper (Boudesocque et al. 2007). Additionally, copper distribution in the soil may not be uniform. As reported by Jacobson et al. (2007), entire regions of vineyard soils were devoid of the metal. Localized hotspots of copper were associated with minimally degraded organic matter, which may have been the result of reduced microbial degradation in those places.

Although the majority of study authors agree that soil organic matter is a very important component in copper speciation, there are some controversies among them on calcareous soils. A few authors have concluded that the main mechanism of copper retention is its precipitation as $CuCO_3$ (Ponizovsky et al. 2007; Rodriguez-Rubio et al. 2003). In contrast, Strawn and Baker (2008) concluded that copper in calcareous soil was predominantly associated with soil organic matter and not with metal oxides, silicates, phosphates, or carbonates.

Soil pH plays an important role in copper retention and mobility. The concentration of uncomplexed copper increases at low pH, thus increasing its mobility (McBride and Bouldin 1984; Temminghoff et al. 1997). At pH 3.9, only 30% of copper was associated with soil organic matter. In comparison, at pH 6.6 the copper–soil organic matter association was 99% (Rodriguez-Rubio et al. 2003; Temminghoff et al. 1997). Thus, uncomplexed, free copper may be more toxic to plants, especially in higher acidity soil.

2.4 Copper Environmental Fate in Air

Copper is found as a trace element in atmospheric water (e.g., fog, clouds, and rain) as a result of its global cycling (Kieber et al. 2004). Atmospheric copper derives from both natural and anthropogenic sources. Natural sources include windblown dust, plant exudates, and sea salt spray. Anthropogenic sources include iron, steel, and non-ferrous metal manufacturing; the burning of fossil fuels; waste incinerators; and terrestrial pesticide use (CDPR 2009b; Hsiao et al. 2002; Kieber et al. 2004). Research results have indicated that the majority of the copper that is released into the atmosphere originates from continental anthropogenic sources.

The primary route by which copper is removed from the atmosphere is wet deposition (Church et al. 1984; Kieber et al. 2004). Kieber et al. (2004) estimated the amounts of Cu per year removed by rain to be 150×10^6 kg. This number also represents the total estimated copper input (continental and marine) into the global atmosphere. Few studies have been performed on the contribution of atmospheric copper to water bodies. Williams et al. (1998) estimated that about 6% of the total copper input into the Irish Sea comes from the atmospheric compartment. Giusti et al. (1993) estimated the atmospheric copper input to the oceans to be $14 - 45 \times 10^6$ kg/year for the dissolved form (e.g., rainwater) and $2-7 \times 10^6$ kg Cu/year as particulates.

In air, copper may exist as several chemical species. In fly ash, generated from municipal solid waste incinerators, copper was found to exist as $CuCO_3$, $Cu(OH)_2$, and CuO (Hsiao et al. 2002). Atmospheric copper is subject to redox reactions and tends to be hydrated. Researchers who investigated the speciation of copper in rainwater reported the volume-weighted average concentrations of TDCu to be

5.3 nM, Cu^+ 1.4 nM, and Cu^{2+} 3.2 nM (Kieber et al. 2004). About half of the TDCu was bound to strong organic ligands and the remainder was bound to weaker organic ligands, inorganic ligands, or existed as free hydrated ions.

3 Effects on Aquatic Organisms

Copper is an essential trace element needed at miniscule levels for the proper functioning of all organisms. However, excessive amounts of copper interfere with vital biological functions. Different species, and even organisms within the same species, can exhibit different sensitivities to elevated copper levels in the water column. Organisms have different mechanisms by which they cope with and process copper. Some organisms bioaccumulate and store copper, whereas others actively regulate its levels. In general, copper is actively regulated in fish, decapod crustaceans, and algae and stored in bivalves, barnacles, and aquatic insects (Brix and Deforest 2000). Therefore, to properly evaluate the copper-related effects on aquatic life, one must understand the factors that affect the biological fate of copper and the mechanisms by which it acts to produce its toxicity.

Copper undergoes complex speciation in natural waters; some species are bioavailable (free Cu^{2+} and Cu^+ ions (Fig. 2)), while others are not. Only bioavailable forms of copper are considered to be toxic to exposed organisms. The reference to "copper" and "free copper" in the following discussion refers to its bioavailable form. The bioavailability, biodistribution to various parts of the organism, and bioaccumulation of copper are dramatically influenced by water chemistry. Therefore, water pH, hardness, organic content, and salinity play important roles in copper-induced toxicity.

The majority of studies in which the toxicity of copper has been addressed were performed on freshwater species. Copper is generally more toxic to organisms in freshwater than in saltwater. One of the reasons for this difference is that freshwater lacks cations, which compete with Cu^{2+} at the biological action sites, thus reducing copper toxicity (Brooks et al. 2007). Consequently, pH and water hardness play more important roles in freshwater than in saltwater environments. Increased pH accentuates copper toxicity because of the reduced competition between copper and hydrogen ions at the cell surface (Wilde et al. 2006). Cations that are involved in water hardness (i.e., Ca^{2+} and Mg^{2+}) also compete with Cu^{2+} for biological binding sites (Boulanger and Nikolaidis 2003). Therefore, Cu^{2+} is less bioavailable in hard water than in soft water.

Although water pH and hardness protect organisms against Cu toxicity to some degree, the DOC content is among the most important factors in reducing copper toxicity to both fresh- and salt-water species. DOC forms organic complexes with copper and thereby reduces copper's bioavailability. According to McIntyre et al. (2008), water hardness and pH are unlikely to protect fish from copper-induced sensory neurotoxicity. However, water that contains high DOC concentrations does diminish the toxic effects of copper on the peripheral olfactory nervous system in

Coho salmon (*Oncorhynchus kisutch*) (McIntyre et al. 2008). High DOC levels also significantly decrease acute copper toxicity to the freshwater flea, *Daphnia magna*, and the estuarine copepod, *Eurytemora affinis* (Hall et al. 2008; Kramer et al. 2004).

Study results show that the water salinity gradient can also significantly affect the biological fate of copper. Water salinity influences the biodistribution and bioaccumulation of copper and can affect its toxicity as well (Amiard-Triquet et al. 1991; Blanchard and Grosell 2005; Grosell et al. 2007; Hall et al. 2008). The biodistribution of copper throughout the gill, intestine, and liver of the common killifish, *Fundulus heteroclitus*, is salinity dependent (Blanchard and Grosell 2005). According to these authors, the gill and the liver are important target organs for copper accumulation at low salinities, whereas the intestine is a target organ at high salinities. In addition, the liver is a major organ involved in copper homeostasis and accumulates the highest amounts of copper. For this reason, the liver may be a potential target organ during chronic copper exposure. Water salinity influences the biodistribution and the toxicity of copper. Grosell et al. (2007) found killifishes to be most tolerant to copper exposure at intermediate salinities, and the acute toxicity was significantly higher in the lowest and highest salinity water. Increased fish sensitivity at both salinity extremes can be attributed to two factors: changes in copper speciation and changes in fish physiology in changing aquatic environments.

In general, water salinity may be more important to species that actively regulate internal osmotic pressure. The majority of invertebrates, however, are osmoconformers. Hence, to them the salinity gradient may be less important. Although in bivalves, the biological fate of copper was salinity dependent, in copepods (*Eurytemora affinis*) the toxicity of copper correlated better to DOC content than water salinity (Hall et al. 2008). In oysters, copper accumulation was inversely related to salinity (Amiard-Triquet et al. 1991). Some species can adapt to tolerate higher pollutant levels. Damiens et al. (2006) described adult oysters that lived in polluted water, wherein their larvae become less sensitive to pollution over time. Phytoplankton species have different sensitivities to copper toxicity: cyanobacteria appear to be most sensitive, coccolithophores and dinoflagellates show intermediate sensitivity, and diatoms are resistant to copper (Brand et al. 1986; Beck et al. 2002).

In many aquatic animals, copper causes toxicity by impairing osmoregulation and ion regulation in the gill (Blanchard and Grosell 2005; McIntyre et al. 2008). When bioavailable Cu^{2+} enters the cell, it is reduced to Cu^+. This copper oxidation state has a high affinity to sulfhydryl groups that are abundant within ATPase enzymes (Katranitsas et al. 2003; Viarengo et al. 1996). The best studied copper toxicity pathways involve the inhibition of ATP-driven pumps and ion channels. Katranitsas et al. (2003) discovered that, in brine shrimp, copper inhibited Na/K ATPase and Mg^{2+} ATPase enzyme activity. Similarly, in the mussel, *Mytilus galloprovincialis*, copper interfered with Ca^{2+} homeostasis in the gill, causing disruptions in Na/K ATPase and Ca^{2+} ATPase (Viarengo et al. 1996). In an in vitro study, Corami et al. (2007) investigated lysosomal activity and found that copper acted at two different sites: the proton pump and Cl^- selective channels. Therefore, copper acts by inhibiting enzymes, ATP-driven pumps, and ion channels, resulting in cell toxicity from disruption of cell homeostasis and leading to changes in internal pH balance, membrane potential, and osmosis.

In addition to inhibiting ATPase enzymes and disrupting ion flow, copper toxicity can be induced by generating reactive oxygen species (ROS) (Bopp et al. 2008; Viarengo et al. 1996). ROS can lead to different outcomes: genotoxicity via DNA strand break and impaired cell membrane permeability via lipid peroxidation. Both pathways compromise normal cell functions.

A less understood effect of copper is neurotoxicity to fish olfaction. There is evidence that exposure to sublethal copper levels results in the loss of chemosensory function, which affects predator-avoidance behavior (McIntyre et al. 2008). The exact mechanisms are not yet completely understood and are still under investigation. Tilton et al. (2008) revealed that copper depresses the transcription of key genes within the olfactory signal transduction pathway.

The environmentally relevant copper levels that interfere with fish chemosensory mechanisms are very low. TDCu concentrations in the range of 0–20 ppb affected sensory capacity and behavior in salmon (Sandahl et al. 2007). At higher levels, copper caused a degeneration of the sensory epithelium (Bettini et al. 2006; Hansen et al. 1999). These effects were observed within hours of exposure. Hence, fish olfaction is a vulnerable endpoint that should be considered in environmental risk assessment.

The developmental stage of fish during their exposure to elevated copper levels may be a critical factor in their sensitivity. Carreau and Pyle (2005) showed that exposure to copper during embryonic development can lead to permanently impaired chemosensory functions. In contrast, fish that are exposed to elevated copper later in life can recover their chemosensory ability after the toxicant is removed.

Copper is stored and transported inside an organism as inorganic and organic complexes. In killifishes, copper bioaccumulates in target organs primarily as copper carbonate ($CuCO_3$) and, to a lower extent, as copper hydroxide ($CuOH^+$ and $Cu(OH)_2$) (Blanchard and Grosell 2005). Bivalves accumulate considerable amounts of copper that is associated with a cytosolic protein called metallothionein (Claisse and Alzieu 1993; Damiens et al. 2006). Although copper bioaccumulates and biodistributes to different organs, it is an internally regulated essential micronutrient. Therefore, according to Brix and Deforest (2000), there is an inverse relationship between metal concentrations in the water and in the organism. Hence, the bioconcentration factor (BCF) is not a suitable concept to describe the bioconcentration of copper.

Toxicity data for aquatic species for copper oxide, selected from the U.S. EPA ECOTOX database, are summarized in Table 2 (U.S. EPA 2009a). The table is divided into sections for freshwater and saltwater organisms. Data are presented for fish, invertebrates, and plants. The toxicity endpoints are also presented in the table, as is the chemical concentration that was lethal (LC_{50}) or produced an effect (EC_{50}). There is a large range in copper toxicity values for different freshwater algae.

Table 2 Copper (I) oxide toxicity to aquatic organisms

Toxicity to freshwater aquatic organisms

Species name: scientific/common	Endpoint	Duration/effects	Concentration $\mu g/L$ (ppb)	Purity (%)
Fish:				
Danio rerio/Zebra danio	LC_{50}	96 h/mortality	75	100
Invertebrates:				
Daphnia similis/Water flea	EC_{50}	48 h/mortality	42	100
Biomphalaria glabrata/Snail	LC_{50}	48 h/mortality	179	100
Plants:				
Pseudokirchneriella subcapitata/Green algae	EC_{50}	30 min, 35 min physiology/photosynthesis	90, 1900, >4500	Not reported
	EC_{50}	96 h/physiology/ photosynthesis	1300, 1600	Not reported
	EC_{50}	96 h/population/ Abundance	30, 60, 230	Not reported

Table 2 (continued)

Toxicity to saltwater aquatic organisms

Species name: scientific/common	Endpoint	Duration/effects	Concentration μg/L (ppb)	Purity (%)
Fish:				
Cyprinodon variegates/Sheepshead minnow	LC_{50}	96 h/mortality	>173	93
Melanogrammus aeglefinus/Haddock	LT_{50}	4.5 h, 5.7 h Mortality	1800	100
Invertebrates:				
Americamysis bahia/Opossum shrimp	LC_{50}	96 h/ mortality	69.7	97
Balanus improvisus/Barnacle	LC_{50}	12 h/mortality	700	Not reported
	LC_{50}	24 h/mortality	500	Not reported
	LC_{50}	48 h/mortality	350	Not reported
	LC_{50}	72 h/mortality	140	Not reported
	LC_{50}	96 h/mortality	20	Not reported

Source: U.S. EPA (2009a) ECOTOX database (accessed: 11/05/09)

4 Nanocopper: Emerging Ecotoxicity Data

A definition of nanotechnology, according to the U.S. EPA White Paper produced in 2007, is "research and technology development at the atomic, molecular, or macro-molecular levels using a length scale of approximately 1–100 nanometers in any dimension; the creation and use of structures, devices and systems that have novel properties and functions because of their small size; and the ability to control or manipulate matter on an atomic scale" (U.S. EPA 2007). A nanometer is one bil-lionth of a meter (10^{-9}); this is equal to the diameter of a single-strand DNA molecule. Indeed, manipulating materials at the molecular and atomic scale pro-duces novel materials that have new physical and chemical properties that may vary from their bulk forms. Rapid growth of the nanotechnology industry and increasing mass production of engineered nanomaterials will inevitably result in environmental exposure to these types of chemicals.

Today, metal nanoparticles are among the most popular types of nanomaterials. Metal nanoparticles like CuO, ZnO, TiO_2, nanosilver, and nanogold have a wide variety of applications, including use in industry, consumer products, medicine, and pesticide products. Copper oxide nanoparticles are used as additives in inks, plastics, lubricants; as coatings for integrated circuits and batteries; and as bactericides for air and liquid filtration (Griffitt et al. 2007; Midander et al. 2009). Thus, metal nanopar-ticles from various sources, including a growing number of pesticide products, could make their way to the surface waters.

Unfortunately, little published information exists on the environmental fate of nanometals, including nanocopper. Metal nanoparticles, when added to the water, can aggregate, sediment out of the water column, adsorb to nutrients, and disasso-ciate to release soluble metal ions (Griffitt et al. 2009; Kahru et al. 2008). Gao et al. (2009) indicated that both water chemistry and the reactivity of the nanoparticle itself should be considered in environmental speciation studies. Hence, laboratory experiments that use deionized water and artificial methods to suspend nanoparticles may not realistically reflect what occurs in natural environments.

The effects of nanocopper on aquatic organisms have not been well studied. Existing studies indicate that copper toxicity strongly depends on particle size. As particle size decreases, toxicity increases. Among the studies that have been performed, there is a 15- to 65-fold increase in toxicity when nano-sized copper particles are used (Table 3). In most studies, the increase in nanocopper toxicity is attributed to an increase in solubility and, consequently, bioavailability (Aruoja et al. 2009; Heinlaan et al. 2008; Mortimer et al. 2010).

However, increased solubility does not always explain increased nanocopper tox-icity. Copper nanoparticles can induce toxicity by mechanisms that are different from those of soluble ions (Griffitt et al. 2007, 2008, 2009). When exposed to equiv-alent bioavailable amounts of nano- and soluble metal-forms, gill copper uptake was identical in zebrafish. However, nanocopper caused greater damage to the gill. Nanocopper produced different morphological effects and global gene expression patterns in the gill than did soluble copper ions alone. Similarly, Kasemets et al. (2009) reported that soluble copper ions explained 50% of nanocopper toxicity in

Table 3 Bulk vs. nanocopper toxicity to different species of aquatic organisms

Species	Duration/effect	Bulk-CuO (mg/L)	n-CuO (mg/L)	Cause of toxicity	Reference
Pseudokirchneriella subcapitata (algae)	72 h EC_{50}	11.55	0.71	Soluble Cu^{2+}	Aruoja et al. (2009)
Vibrio fischeri (bacteria)	1/2 h EC_{50}	3899	79	Soluble Cu^{2+}	Heinlaan et al. (2008)
Daphnia magna (crustacean)	48 h EC_{50}	164.8	3.2	Soluble Cu^{2+}	Heinlaan et al. (2008)
Thamnocephalus platyurus (crustacean)	24 h EC_{50}	94.5	2.1	Soluble Cu^{2+}	Heinlaan et al. (2008)
Saccharomyces cerevisiae	8 h EC_{50}	1297	20.7	50% soluble Cu^{2+}	Kasemets et al. (2009)
(yeast)	24 h EC_{50}	873	13.4		
Tetrahymena thermophila	4 h EC_{50}	1705	128	Soluble Cu^{2+}	Mortimer et al. (2010)
(protozoa)	24 h EC_{50}	1966	97.9		

n = nano

yeast. In vitro studies provided evidence to show that copper nanoparticles have the ability to cause mitochondrial (Karlsson et al. 2009) and DNA damage (Midander et al. 2009). Although the mechanisms of nanoparticle toxicity are not well understood, the findings to date suggest that both ionic copper and nanoparticulate copper are responsible for the toxicity that is produced.

5 Monitoring and Ambient Water Quality Standards

Because of the heavy use of copper oxide-based boat antifouling paints in poorly flushed marine environments, copper monitoring data in marinas are particularly useful for assessing potential water quality impacts. Several copper monitoring studies have been conducted in California marinas (Table 4). Singhasemanon et al. (2009) reported median DCu concentrations from 23 marinas that reflected a range of water salinities. Different water types were determined by measuring water electrical conductivity (EC) in micro-Siemens/centimeter (μS/cm). The values are 0–1,500 for freshwater, 1,500–15,000 for brackish water, and >15,000 for saltwater. "There were significant differences in median DCu concentrations among the three water types (one-way ANOVA, $F_{2, 64} = 8.90$, $p < 0.0005$), with freshwater marina median DCu concentrations being significantly less than those in salt and brackish

Table 4 Dissolved copper concentrations (μg/L) measured in the water column of California marinas at different water salinities and the CTR standard values (μg/L)

Source	Saltwater	Brackish water	Freshwater
	DCu	DCu	DCu
Singhasemanon et al. (2009)	3.7 (median)	2.6	1.4
RWQCB (2007)	4.27 (mean)	–	–
Schiff et al. (2007)	7.0 (mean)	–	–
CTR[a]	3.1/4.8	3.1/4.8	9.0/13[b]

Source: U.S. EPA (2000)
(CTR- California Toxics Rule; CCC- criterion continuous concentration; CMC- criterion maximum concentration; DCu- dissolved copper)
[a]CCC and CMC, respectively
[b]The CTR values for freshwater are based on total water hardness (100 mg/L) and will change depending on the individual fresh water body

water marinas (Tukey's Test, family error rate = 0.05). In contrast, there was no significant difference between median DCu concentrations in salt and brackish water marinas" (Singhasemanon et al. 2009). This suggests that there are similarities in the sources of dissolved copper or in the physical and chemical processes that are driving the cycling of dissolved copper in saltwater and brackish water marinas.

Median DCu concentrations of marina samples were greater than median concentrations found at their associated local reference sites (LRSs) (Singhasemanon et al. 2009). Among the three water types, median DCu concentrations were 3.7, 2.6, and 1.4 μg/L for saltwater, brackish water, and freshwater marinas, respectively. For comparison, the associated LRS median concentrations were 0.6, 1.6, and 0.5 μg/L. Through source analysis, Singhasemanon et al. (2009) concluded that during the dry weather season (July through October), antifouling paints are probably a major source of copper pollution in California saltwater and brackish water marinas. Similar data from studies performed in Southern California also indicated elevated DCu concentrations in saltwater marinas during the dry season (Table 4) (RWQCB 2007; Schiff et al. 2007). The authors of a study conducted by the Santa Ana RWQCB (2007) further concluded that DCu from copper-containing boat coatings may be settling in marina sediments of Lower Newport Bay.

California Toxics Rule (CTR) standards pertain to the chemical concentrations in inland surface waters and enclosed bays and estuaries and are intended to protect human health and the environment (U.S. EPA 2009b). CTR standards establish freshwater and saltwater thresholds for chemicals, based on criterion continuous concentrations (CCC) for chronic toxicity endpoints and criterion maximum concentrations (CMC) for acute toxicity endpoints (Table 4). In their study, Singhasemanon et al. (2009) found that 51 and 30% of their brackish water and saltwater marina samples exceeded the CTR's CCC and CMC standards, respectively. In contrast, none of their freshwater marina samples exceeded the CTR standards. This suggests that elevated copper concentrations in some saltwater and brackish water marinas may pose a risk to aquatic life.

6 Summary

Besides being a naturally occurring element and an essential micronutrient, copper is used as a pesticide, but at generally higher concentrations. Copper, unlike organic pesticides, does not degrade, but rather enters a complex biogeochemical cycle. In the water column, copper can exist bound to both organic and inorganic species and as free or hydrated copper ions. Water column chemistry affects copper speciation and bioavailability. In all water types (saltwater, brackish water, and freshwater), organic ligands in the water column can sequester the majority of dissolved copper, and therefore, organic ligands play the largest role in copper bioavailability. In freshwater, however, the geochemistry of a particular location, including water column characteristics such as water hardness and pH, is a significant factor that can increase copper bioavailability and toxicity. In most cases, organic ligand concentrations greatly exceed copper ion concentrations in the water column and therefore provide a large buffering capacity. Hence, copper bioavailability can be grossly overestimated if it is based on total dissolved copper (TDCu) concentrations alone. Other factors that influence copper concentrations include location in the water column, season, temperature, depth, and level of dissolved oxygen. For example, concentrations of bioavailable copper may be significantly higher in the bottom waters and sediment pore waters, where organic ligands degrade much faster and dissolved copper is constantly resuspended and recycled into the aquatic system.

Aquatic species differ greatly in their sensitivity to copper. Some animals, like mollusks, can tolerate high concentrations of the metal, while others are adversely affected by very low concentrations of copper. Emerging evidence shows that very low, sublethal copper levels can adversely affect the sense of smell and behavior of fish. The developmental stage of the fish at the time of copper exposure is critical to the reversibility of sensory function effects. The fish olfactory system may be the most sensitive structure to copper pollution.

The major factors that influence copper-induced toxicity are dissolved organic carbon and water salinity. Dissolved organic carbon reduces copper toxicity by sequestering bioavailable copper and forming organic complexes with it. Salinity, on the other hand, influences copper bioavailability at the biological action site and also affects metal biodistribution and bioaccumulation in the organism. Therefore, the salinity gradient can increase or decrease copper toxicity in different aquatic species. In some killifish, copper may affect different organs at different times, depending on the water salinity.

The most studied and best explained copper toxicity mechanisms involve inhibition of key enzymes and disruption of osmoregulation in the gill. Other toxicity mechanisms may involve reactive oxygen species generation and changes of gene transcription in the fish olfactory signaling pathway.

More studies are needed to evaluate the potential magnitude of copper remobilization from the sediment that may result from climate change and its effects on surface waters. Moreover, the environmental exposure, fate, and ecotoxicity of emerging metal nanoparticles, including nanocopper, will require additional studies as new forms of copper appear from application of nanotechnology to copper compounds.

Acknowledgments A special thanks to K. Goh and D. Oros from the Environmental Monitoring Branch of the California Environmental Protection Agency, Department of Pesticide Regulation, for their help in preparing this chapter.

References

Amiard-Triquet C, Berthet B, Martoja R (1991) Influence of salinity on trace metal (Cu, Zn, Ag) accumulation at the molecular, cellular and organism level in the oyster *Crassostrea gigas* Thunberg. Biol Met 4:144–150

Apte SC, Gardner MJ, Ravenscroft JE (1990) An investigation of copper complexation in the Severn estuary using differential pulse cathodic stripping voltammetry. Mar Chem 29: 63–75

Aruoja V, Dubourguier HC, Kasemets K, Kahru A (2009) Toxicity of nanoparticles of CuO, ZnO and TiO$_2$ to microalgae *Pseudokirchneriella subcapitata*. Sci Total Environ 407:1461–1468

Beck NG, Bruland KW, Rue EL (2002) Short-term biogeochemical influence of a diatom bloom on the nutrient and trace metal concentrations in South San Francisco Bay microcosm experiments. Estuaries 25(6A):1063–1076

Beck AJ, Sanudo–Wilhelmy SA (2007) Impact of water temperature and dissolved oxygen on copper cycling in an urban estuary. Environ Sci Technol 41(17):6103–6108

Berner RA (1980) Early diagenesis, a theoretical approach. Princeton Series in Geochemistry. Princeton University press, Princeton, NJ

Bettini S, Ciani F, Franceschini V (2006) Recovery of the olfactory receptor neurons in the African *Tilapia mariae* following exposure to low copper level. Aquat Toxicol 76:321–328

Blanchard J, Grosell M (2005) Effects of salinity on copper accumulation in the common killifish (*Fundulus heteroclitus*. Environ Toxicol Chem 24(6):1403–1413

Bopp SK, Abicht HK, Knauer K (2008) Copper- induced oxidative stress in rainbow trout gill cells. Aquat Toxicol 86:197–204

Boudesocque S, Guillon E, Aplincourt M, Marceau E, Stievano L (2007) Sorption of Cu(II) onto vineyard soils: macroscopic and spectroscopic investigations. J Colloid Interface Sci 307: 40–49

Boulanger B, Nikolaidis NP (2003) Mobility and aquatic toxicity of copper in an urban watershed. J Am Water Resour Assoc 39(2):325–336

Brand L, Sunda WG, Guillard RRL (1986) Reduction of marine phytoplankton reproduction rates by copper and cadmium. J Exp Mar Biol Ecol 96:225–250

Brix KV, DeForest DK (2000) Critical review of the use of bioconcentration factors for hazard classification of metals and metal compounds. Parametrix Inc., Washington, DC. Report No. 555-3690-001

Brooks SJ, Bolam T, Tolhurst L, Bassett J, Roche JL, Waldock M, Barry J, Thomas KV (2007) Effects of dissolved organic carbon on the toxicity of copper to the developing embryos of the pacific oyster (*Crassostrea gigas*). Environ Toxicol Chem 26(8):1756–1763

Buck KN, Bruland KW (2005) Copper speciation in San Francisco Bay: a novel approach using multiple analytical windows. Mar Chem 96(1–2):185–198

Buck KN, Ross JRM, Flegal AR, Bruland KW (2007) A review of total dissolved copper and its chemical speciation in San Francisco Bay, California. Environ Res 105:5–19

Buckley PJM, Van Den Berg CMG (1986) Copper complexation profiles in the Atlantic Ocean. Mar Chem 19:281–296

Carreau ND, Pyle GG (2005) Effect of copper exposure during embryonic development on chemosensory function of juvenile fathead minnows (*Pimephales promelas*). Ecotoxicol Environ Saf 61:1–6

CDPR (1991) Cuprous Oxide, Grade AA. EPA Reg No 63005-1, section 3 registration, 50339-042. 1220 N Street, Sacramento, CA 95814, USA

CDPR (2009a) (California Department of Pesticide Regulation) CDPR Database. URL:http://apps.cdpr.ca.gov/cgi-bin/label/labq.pl?p_chem=175&activeonly=on. Accessed 4 Nov 2009

CDPR (2009b) (PUR, Pesticide Use Reporting Database) URL: http://www.cdpr.ca.gov/docs/pur/purmain.htm. Accessed 9 Nov 2009

Chester R, Thomas A, Lin FJ, Basaham AS, Jacinto G (1988) The solid state speciation of copper in surface water particulates and oceanic sediments. Mar Chem 24:261–292

Church TM, Tramontano JM, Scudlark JR, Jickells TD, Tokos JJ Jr, Knap AH, Galloway JN (1984) The wet deposition of trace metals to the Western Atlantic Ocean at the mid-Atlantic coast and on Bermuda. Atmos Environ 18(12):2657–2664

Claisse D, Alzieu C (1993) Copper contamination as a result of antifouling paint regulations? Mar Pollut Bull 26(7):395–397

Corami F, Capodaglio G, Turetta C, Bragadin M, Calace N, Petronio BM (2007) Complexation of cadmium and copper by fluvial humic matter and effects on their toxicity. Annal Chim (Italy) 97:25–37

Damiens G, Mouneyrac C, Quiniou F, His E, Gnassia-Barelli M, Romeo M (2006) Metal bioaccumulation and metallothionein concentrations in larvae of *Crassostrea gigas*. Environ Pollut 140:492–499

Gao J, Youn S, Hovsepyan A, Llaneza VL, Wang Y, Bitton G, Bonzongo JC (2009) Dispersion and toxicity of selected manufactured nanomaterials in natural river water samples: effects of water chemical composition. Environ Sci Technol 43(9):3322–3328

Georgopoulos PG, Roy A, Yonone-Lioy MJ, Opiekun RE, Lioy PJ (2001) Copper: environmental dynamics and human exposure issues. J Toxicol Environ Health, Part B 4:341–394

Gerringa LJA (1990) Aerobic degradation of organic matter and the mobility of Cu, Cd, Ni, Pb, Zn, Fe and Mn in the marine sediment slurries. Mar Chem 29:355–374

Gerringa LJA, Van Den Meer J, Cauwet G (1991) Complexation of copper and nickel in the dissolved phase of marine sediment slurries. Mar Chem 36:51–70

Giusti L, Yang YL, Hewitt CN, Hamilton-Taylor J, Davison W (1993) The solubility and partitioning of atmospherically derived trace metals in artificial and natural waters: a review. Atmos Environ 27A(10):1567–1578

Goh KS (1987) Cuprous oxide. In: Worthing CR et al (eds) The pesticide manual: a world compendium, 8th edn. The British Crop Protection Council, UK, pp 196–197

Goh BPL, Chou ML (1997) Heavy metal levels in marine sediments of Singapore. Environ Monit Assess 44:67–80

Griffitt RJ, Hyndman K, Denslow ND, Barber DS (2009) Comparison of molecular and histological changes in zebrafish gills exposed to metallic nanoparticles. Toxicol Sci 107(2):404–415

Griffitt RJ, Luo J, Gao J, Bonzongo JC, Barber DS (2008) Effects of particle composition and species on toxicity of metallic nanomaterials in aquatic organisms. Environ Toxicol Chem 27(9):1972–1978

Griffitt RJ, Weil R, Hyndman KA, Denslow ND, Powers K, Taylor D, Barber DS (2007) Exposure to copper nanoparticles causes gill injury and acute lethality in zebrafish (*Danio rerio*). Environ Sci Technol 41(23):8178–8186

Grosell M, Blanchard J, Brix KV, Gerdes R (2007) Physiology is pivotal for interactions between salinity and acute copper toxicity to fish and invertebrates. Aquat Toxicol 84:162–172

Gundersen P, Steinnes E (2003) Influence of pH and TOC concentration on Cu, Zn, Cd, and Al speciation in rivers. Water Res 37:307–318

Hall LW Jr, Anderson RD, Lewis BL, Arnold WR (2008) The influence of salinity and dissolved organic carbon on the toxicity of copper to the estuarine copepod, *Eurytemora affinis*. Arch Environ Contam Toxicol 54:44–56

Hansen JA, Rose JD, Jenkins RA, Gerow KG, Berbmah HL (1999) Chinook salmon (*Oncorhynchus tshawytscha*) and rainbow trout (*Oncorhynchus mykiss*) exposed to copper: neurophysiological and histological effects on the olfactory system. Environ Toxicol Chem 18:1979–1991

Heinlaan M, Ivask A, Blinova I, Dubourguier HC, Kahru A (2008) Toxicity of nanosized and bulk ZnO, CuO and TiO$_2$ to bacteria *Vibrio fischeri* and crustaceans *Daphnia magna* and *Thamnocephalus platyurus*. Chemosphere 71:1308–1316

Helland A, Bakke T (2002) Transport and sedimentation of Cu in a microtidal estuary, SE Norway. Mar Pollut Bull 44:149–155

Hirose K, Dokiya Y, Sugimura Y (1982) Determination of conditional stability constants of organic copper and zinc complexes dissolved in seawater using ligand exchange method with EDTA. Mar Chem 11:343–354

Hoffman SR, Shafer MM, Armstrong DE (2007) Strong colloidal and dissolved organic ligands binding copper and zinc in rivers. Environ Sci Technol 41(20):6996–7002

HSDB (Hazardous Substances Data Bank) (2008) U.S. National Library of Medicine. URL: http://toxnet.nlm.nih.gov/cgi-bin/sis/htmlgen?HSDB. Accessed 8 July 2008

Hsiao MC, Wang HP, Wei YL, Chang J, Jou CJ (2002) Speciation of copper in the incineration fly ash of a municipal solid waste. J Hazard Mater B 91:301–307

Hurst MP, Bruland KW (2005) The use of nafion-coated thin mercury film electrodes for the determination of the dissolved copper speciation in estuarine water. Anal Chim Acta 546:68–78

ILO (International Labor Organization) (2008) International Occupational Safety and Health Information Center. ICSC: 0421-Copper(I)oxide. URL: http://www.ilo.org/legacy/english/protection/safework/cis/products/icsc/dtasht/_icsc04/icsc0421.htm. Last updated 01/29/2008

Jacobson AR, Dousset S, Andreux F, Baveye PC (2007) Electron microprobe and synchrotron x-ray fluorescence mapping of the heterogeneous distribution of copper in high-copper vineyard soils. Environ Sci Technol 41(18):6343–6349

Jones B, Bolam T (2007) Copper speciation survey from UK marinas, harbors and estuaries. Mar Pollut Bull 54:1127–1138

Kabala C, Singh BR (2001) Fractionation and mobility of copper, lead and zinc in soil profiles in the vicinity of a copper smelter. J Environ Qual 30:485–492

Kahru A, Dubourguier HC, Blinova I, Ivask A, Kasemets K (2008) Biotests and biosensors for ecotoxicology of metal oxide nanoparticles: a minireview. Sensors 8:5153–5170

Karlsson HL, Gustafsson J, Cronholm P, Möller L (2009) Size-dependent toxicity of metal oxide particles- a comparison between nano- and micrometer size. Toxicol Lett 188:112–118

Kasemets K, Ivask A, Dubourguier HC, Kahru A (2009) Toxicity of nanoparticles of ZnO, CuO and TiO$_2$ to yeast *Saccharomyces cerevisiae*. Toxicol In Vitro 23:1116–1122

Katranitsas A, Castritsi-Catharios J, Persoone G (2003) The effects of a copper-based antifouling paint on mortality and enzymatic activity of a non-target marine organism. Mar Pollut Bull 46:1491–1494

Kieber RJ, Skrabal SA, Smith C, Willey JD (2004) Redox speciation of copper in rainwater: temporal variability and atmospheric deposition. Environ Sci Technol 38(13):3587–3594

Klinkhammer GP (1980) Early diagenesis in sediments from the Eastern Equatorial Pacific, II. Pore water metal results. Earth Planet Sci Lett 49:81–101

Klinkhammer G, Heggie DT, Graham DW (1982) Metal diagenesis in oxic marine sediments. Earth Planet Sci Lett 61:211–219

Kozelka PB, Bruland KW (1998) Chemical speciation of dissolved Cu, Zn, Cd, Pb in Narragansett Bay, Rhode Island. Mar Chem 60:267–282

Kramer KJM, Jak RG, Van Hattum B, Hoftman RN, Zwolsman JJG (2004) Copper toxicity in relation to surface water-dissolved organic matter: biological effects to *Daphnia magna*. Environ Toxicol Chem 23(12):2971–2980

Landing WM, Feely RA (1981) The chemistry and vertical flux of particles in the Northeastern Gulf of Alaska. Deep Sea Res Part A 28:19–37

Liu SH, Wang HP (2004) *In situ* speciation studies of copper-humic substances in a contaminated soil during electrokinetic remediation. J Environ Qual 33:1280–1287

Long KWJ (2006) Copper: marine fate and effects assessment. CSI Europe. Pentlands Science Park, project number 06741. Penicuik, Midlothian EH26 0PZ, UK

Lucia M, Campos AM, Van Den Berg CMG (1994) Determination of copper complexation in sea water by cathodic stripping voltammetry and ligand competition with salicylaldoxime. Anal Chim Acta 284:481–496

McBride MB, Bouldin DR (1984) Long-term reactions of copper (II) in a contaminated calcareous soil. Soil Sci Soc Am J 48:56–59

McIntyre JK, Baldwin DH, Meador JP, Scholz NL (2008) Chemosensory deprivation in juvenile Coho salmon exposed to dissolved copper under varying water chemistry conditions. Environ Sci Technol 42(4):1352–1358

Midander K, Cronholm P, Karlsson HL, Elihn K, Möller L, Leygraf C, Wallinder IO (2009) Surface characteristics, copper release, and toxicity of nano- and micrometer-sized copper and copper(II) oxide particles: a cross-disciplinary study. Small 5(3):389–399

Moffett JW, Zilka RG (1983) Oxidation kinetics of Cu (I) in seawater: implications for its existence in the marine environment. Mar Chem 13:239–251

Mortimer M, Kasemets K, Kahru A (2010) Toxicity of ZnO and CuO nanoparticles to ciliated protozoa *Tetrahymena thermophila*. Toxicology 269:182–189

Muller FLL (1996) Interactions of copper, lead and cadmium with the dissolved, colloidal and particulate components of estuarine and coastal waters. Mar Chem 52:245–268

Pesavento M, Biesuz R, Profumo A, Soldi T (2003) Investigation of the complexation of metal-ions by strong ligands in fresh and marine water. Environ Sci Pollut Res 10(5):317–320

Ponizovsky AA, Allen HE, Ackerman AJ (2007) Copper activity in soil solutions of calcareous soils. Environ Pollut 145:1–6

Rodriguez-Rubio P, Morillo E, Madrid L, Undabeytia T, Maqueda C (2003) Retention of copper by a calcareous soil and its textural fractions: influence of amendment with two agroindustrial residues. Eur J Soil Sci 54:401–409

RWQCB (Regional Water Quality Control Board) (2007) Santa Ana. California Environmental Protection Agency. (2007) Lower Newport Bay Copper/Metals Marina Study. Final report. URL: http://www.swrcb.ca.gov/rwqcb8/water_issues/programs/tmdl/docs/newport/finalcufinal_report.pdf. Accessed 22 Feb 2010

Sandahl JF, Baldwin DH, Jenkins JJ, Scholz NL (2007) A sensory system at the interface between urban stormwater runoff and salmon survival. Environ Sci Technol 41:2998–3004

Schiff K, Brown J, Diehl D, Greenstein D (2007) Extent and magnitude of copper contamination in marinas of the San Diego region, California, USA. Mar Pollut Bull 54:322–328

Shank GC, Skrabal SA, Whitehead RF, Kieber RJ (2004) Strong copper complexation in an organic-rich estuary: the importance of allochthonous dissolved organic matter. Mar Chem 88:21–39

Singhasemanon N, Pyatt E, Bacey J (2009) Monitoring for indicators of antifouling paint pollution in California marinas. California Environmental Protection Agency, Department of Pesticide Regulation, Environmental Monitoring Branch, EH08-05. URL: http://www.cdpr.ca.gov/docs/emon/pubs/ehapreps/eh0805.pdf. Accessed 22 Feb 2010

Skrabal SA, Donat JR, Burdige DJ (2000) Pore water distribution of dissolved copper and copper-complexing ligands in estuarine and coastal marine sediments. Geochim Cosmochim Acta 64(11):1843–1857

Snoeyink VL, Jenkins D (1980a) Alkalinity and acidity. In: Water chemistry. Wiley, New York, NY, p 173

Snoeyink VL, Jenkins D (1980b) Complexes with other inorganic ligands. In: Water chemistry. Wiley, New York, NY, p 221

Strawn DG, Baker LL (2008) Speciation of Cu in a contaminated agricultural soil measured by XAFS, μ-XAFS and μ-XRF. Environ Sci Technol 42(1):37–42

Suda WG, Hanson AK (1987) Measurement of free cupric ion concentration in seawater by a ligand competition technique involving copper sorption onto C_{18} SEP-PAK cartridges. Limnol Oceanogr 32(3):537–511

Temminghoff EJM, Van Den Zee SE, De Haan FAM (1997) Copper mobility in a copper-contaminated sandy soil a affected by pH and solid and dissolved organic matter. Environ Sci Technol 31(4):1109–1115

Tilton F, Tilton SC, Bammler TK, Beyer R, Farin F, Stapleton PL, Gallagher EP (2008) Transcription biomarkers and mechanisms of copper-induced olfactory injury in zebrafish. Environ Sci Technol 42(24):9404–9411

U.S. DHHS (Department of Health and Human Services) (2004) Agency for Toxic Substances and Disease Registry (ATSDR) toxicological profile for copper. URL: http://www.atsdr.cdc.gov/toxprofiles/tp132.pdf. Last updated 10/01/2007

U.S. EPA (2007) Office of the science advisor, Nanotechnology white paper. Washington, DC. URL: http://www.epa.gov/osa/nanotech.htm. Last updated 11/04/2009

U.S. EPA (2009a) ECOTOX database. URL: http://cfpub.epa.gov/ecotox/. Accessed 5 Nov 2009

U.S. EPA (2009b) Water quality standards. URL: http://www.epa.gov/waterscience/standards/rules/ctr/. Last updated 03/26/2009

U.S. EPA (Environmental Protection Agency) (2000) Water quality standards; establishment of numeric criteria for priority toxic pollutants for the state of California. 40 CFR part 131 (2000). URL: http://www.epa.gov/fedrgstr/EPA-WATER/2000/May/Day-18/w11106. pdf. Accessed 18 Nov 2009

Van Den Berg CMG (1984) Organic and inorganic speciation of copper in the Irish Sea. Mar Chem 14:201–212

Van Den Berg CMG, Donat JR (1992) Determination and data evaluation of copper complexation by organic ligands in sea water using cathodic stripping voltammetry at varying detection windows. Anal Chim Acta 257:281–291

Viarengo A, Pertica M, Mancinelli G, Burlando B, Canesi L, Orunesu M (1996) In vivo effects of copper on calcium homeostasis mechanisms of mussel gill cell plasma membranes. Comp Biochem Physiol Part C 113(3):421–425

Wang R, Chakrabarti CL (2008) Copper speciation by competing ligand exchange method using differential pulse anodic stripping voltammetry with ethylenediaminetetraacetic acid (EDTA) as competing ligand. Anal Chim Acta 614:153–160

Web Elements (2009) The periodic table on the web. URL: http://www.webelements.com/compounds/copper/dicopper_oxide.html. Accessed 4 Nov 2009

Weser U, Schubotz LM, Younes M (1979) Chemistry of copper proteins and enzymes. Nriagu JO (ed) Copper in the environment. Part. II: health effects. Wiley, Toronto, ON, pp 197–240

Widerlund A (1996) Early diagenetic remobilization of copper in near-shore marine sediments: a quantitative pore-water model. Mar Chem 54:41–53

Wilde KL, Stauber JL, Markich SJ, Franklin NM, Brown PL (2006) The effects of pH on the uptake and toxicity of copper and zinc in a tropical freshwater alga (Chlorella sp). Arch Environ Contam Toxicol 51:174–185

Williams MR, Millward GE, Nimmo M, Fones G (1998) Fluxes of Cu, Pb and Mn to the North-Eastern Irish Sea: the importance of sedimental and atmospheric inputs. Mar Pollut Bull 36(5):366–375

Human Exposure, Biomarkers, and Fate of Organotins in the Environment

**Hussein K. Okoro, Olalekan S. Fatoki, Folahan A. Adekola,
Bhekumusa J. Ximba, Reinette G. Snyman, and Beatrice Opeolu**

Contents

H.K. Okoro (✉)
Department of Chemistry, Faculty of Applied Sciences, Cape Peninsula University
of Technology, Cape Town 7535, South Africa
e-mail: okoroowo@yahoo.com; okorohk@cput.ac.za

D.M. Whitacre (ed.), *Reviews of Environmental Contamination and Toxicology*,
Reviews of Environmental Contamination and Toxicology 213,
DOI 10.1007/978-1-4419-9860-6_2, © Springer Science+Business Media, LLC 2011

1 Introduction

Organotin compounds (OTCs) are organic derivatives of tin (Sn^{4+}) and are characterized by the presence of covalent bonds between three carbon atoms and a tin atom. The organotins are designated as mono-, di-, tri-, or tetra-organotin compounds and have the general formula (n-C_4H_9), Sn–X, where X is an anion or a group linked covalently through a hetero-atom (Dubey and Roy 2003; Okoro et al. 2011). Organotin pollution in the aquatic environment is of global concern; two triorganotin compound groups, the tributyltins and triphenyltins, are toxic to aquatic life (Fent 1996) and are used worldwide not only as biocides in antifouling paints but also as preserving agents for wood and timber, and as agricultural fungicides. These uses result in direct release to water, with consequential uptake and accumulation in aquatic fauna (Harino et al. 2000).

Because the organotins are used as antifouling agents in boat paints, they are common contaminants of marine and freshwater ecosystems. Fent and Muller (1991) detected concentrations of selected organotin species in a wastewater treatment plant in Zurich, Switzerland. It was discovered that municipal wastewater and sewage sludge contain considerable amounts of organotin species [tributyltin (TBT), butyltins (BTs), dibutyltins (DBTs), and monobutyltins (MBTs)]. MBT and DBT occurred as degradation products of TBT, and they are known to have entered the treatment plant as a contaminant of municipal wastewater. Moreover, the leaching and weathering of polyvinyl chloride (PVC) materials that contain OTCs may also result in their release on a large scale (Becker et al. 1997).

Organotin first became a topic of broad interest when it was discovered that antifouling paints were causing the decline of coastal marine mollusks. Such reports first surfaced in the 1970s when the phenomenon of imposex was reported for *Nucella lapillus* in the UK (Blanca 2008). As awareness of the effects of TBT has grown, global efforts to address the problem have increased, and measures have been taken by authorities to protect the aquatic environment from organotins. Hence, the use of TBT on small boats was prohibited by many countries beginning in the mid-1980s (Konstantious and Albanis 2004).

Because detection of environmental contaminants is so critical to their regulation, many methods have been developed to analyze for the OTCs in environmental media (Morabito and Quevauviller 2002). The most successful methods are those that involve separation of TBT and its degradation products by gas chromatography (GC); GC is sensitive and has both high resolving power and selective detection when coupled with mass spectrometry (Delucchi et al. 2007). Sentosa et al. (2009) used an ion-pair reversed-phase chromatography (IR-RP) technique to analyze for speciation of DBT, TBT, and triphenyltin (TPT). These three species were successfully resolved using an ion-pair reversed-phase chromatography column. The eluates were detected online by using a hydride generation-quartz furnace atomic absorption spectrometry (HG-QFAAS) method. The eluent consisted of a mixture of methanol, water, and acetic acid that had a composition of 80:19:1 and contained 1.0 mol L^{-1} of decane sulfonate acid as the ion pairing reagent. The pH of the eluent was adjusted to 1.0 mol L^{-1} H_2SO_4. All species were successfully resolved under these conditions. The capacity factors (k^1) of DBT, TBT, and TPT were 0.27, 2.54, and 5.92, respectively. The resolution (R_s) values of DBT–TBT and TBT–TPT were 9.76 and 3.50, respectively. These values demonstrate the effectiveness of this chromatographic system to resolve the OTCs.

Aquatic organisms exposed to the OTCs have shown various effects. In many marine species, such effects include larval mortality (Bella et al. 2005a) and impairment in growth, development, reproduction, and survival (Haggera et al. 2005). Moreover, the results of several experiments have indicated that there is or may be a spectrum of potential adverse chronic systemic effects of organotin exposure in animals and humans. The type of damage that has been sustained by exposure to organotins in animal testing includes immunosuppression, endocrine effects, neurotoxic effects, and effects on enzymatic activity. In addition to being bioaccumulative, exposure to organotins may also produce the following types of damage: ocular, dermal, cardiovascular, pulmonary, gastrointestinal, blood dyscrasias, reproductive developmental, liver, kidney, and possibly carcinogenic effects (WHO-IPCS 1999; EU-SCOOP 2006; Nakanish 2007).

Although the fate and chemical characteristics of the organotin compounds have been much investigated in developed countries, only limited data are available from Africa. The aim of this chapter is to review the distribution, fate, and measurement of organotins in the environment.

2 Routes of Human Exposure to the Organotins

The OTCs constitute a large class of compounds that have widely varying properties and that have been used for many purposes. The global production, in 2003, was approximately 40,000 t (EVISA 2010). Annual production at such levels, the wide spread use of the OTCs, and their high stability in marine water have led to their presence as contaminants in various ecosystems.

Consumption of contaminated drinking water, beverages, and, in particular, marine food is an important route of human exposure to TBT (Forsyth and Jay 1997; Azuela and Vasconcelos 2002; Chieu et al. 2002). Marine fishery products have been reported to contain high concentrations of OTCs. Therefore, the human diet is expected to have some amounts of the OTCs that will result in human tissue and blood residues (Lo et al. 2003; EFSA 2004; ATSDR 2005; EU-SCOOP 2006). Recent results have shown that fish and fish products are generally the main source of OTCs in the diet; OTCs were detected in whole blood samples of fishermen and their family members, and an association existed of the levels found with age, gender, and level of fish consumption (Pann et al. 2008). These researchers concluded that their results give strong support to the hypothesis that fish constitute the main source of TPT for humans in Finland.

Sadiki and Williams (1999) analyzed Canadian drinking water samples that had been distributed through PVC (polyvinylchloride) pipes. These authors confirmed the presence of OTCs in some drinking water samples collected from residential houses and commercial buildings that were supplied by recently installed PVC piping. The contamination levels detected ranged up to 291 ng (Sn) L^{-1} MMT (monomethyltin trichloride), 49.1 ng (Sn) L^{-1} DMT, 28.5 ng (Sn) L^{-1} MBT, and 52.3 ng L^{-1} (Sn) DBT (dimethyltin dichloride).

Takahashi et al. (1999) reported that several household commodities composed of polymethane, plastic polymers, and silicones, such as diaper covers, sanitary napkins, certain brands of gloves, cellophane wrap, sponges, and baking parchments, contained amounts (up to the $\mu g \ g^{-1}$ level) of several organotin compounds. DBT was detected in treated turkey livers at levels between <0.2 and 6 $\mu g \ g^{-1}$ when DBT derivatives were used as an anthelmintic and coccidiostat in poultry production (Tsuda et al. 1995).

In the UK, a survey showed that organotin levels were generally low in commercial species sampled from many locations throughout the country, and it was suggested that levels found did not present a health risk (FSA 2005). Lo et al. (2003) conducted a study in Germany using eight human volunteers (4 males and 4 females aged 18–54). The serum of the tested individuals exhibited levels of organotin that were below the limits of detection, and TBT and TPT were found at concentration ranges between 0.02–0.05 and 0.17–0.67 $\mu g \ L^{-1}$, respectively. Alzieu (2000) reported that contact exposure to TBT causes irritation of the eyes and skin, potentially leading to severe dermatitis. Because of these properties, it is difficult to guarantee a safe environmental level for TBT. Therefore, use of TBT as a biocide in aquatic systems may well be incompatible with the protection of the ecosystem and with certain marine activities such as oyster farming.

3 Distribution of the Organotins in the Environment

Because of the extensive use of organotins in numerous human activities, large amounts of the OTCs have been introduced to various ecosystems (Blunden and Evans 1990). Significant concentrations of the organotins and their metabolites have been detected in all phases of the aquatic environment: water, suspended matter, sediments, and biomass. The levels of organotins detected in the atmosphere are very low (Blunden and Evans 1990). Among the OTCs, even trace levels of TBT in the environment may be of concern, because it has been considered among the most hazardous compounds to marine organisms (Wagner 1993; Maguire 1996).

3.1 Organotin in Aquatic Systems

OTCs are of concern because of their high toxicity, widespread use, direct input into the environment, and their relatively high persistence. The OTCs enter the aquatic system by many routes. To date, organotin research has been restricted mainly to regions having high shipping volumes, harbors, and/or shipyards, because the primary way in which organotins reach the environment is through use as antifouling agents. TBT in antifouling paints is directly emitted into water, resulting in contaminated water, marine sediments, lakes, and coastal areas (Hoch 2001). As expected, the butyltins have also been detected as residues in marine mammals.

The concentrations of hepatic butyltin reported in fireless porpoise, collected from the Seto Inland Sea, Japan, were as high as 10,000 ng g^{-1} wet wt (wwt), whereas the levels in crustaceans taken from the Japanese coastline ranged from 110 to 5200 ng g^{-1} wwt (Tanabe et al. 1998). Evidence exists to show that legislation introduced to govern the use of TBT in antifouling paints has reduced aquatic concentrations of this contaminant (Fent and Hunn 1995; Dowson et al. 1993).

3.2 Organotin in Sediments

Triorganotin compounds have low aqueous solubility and low mobility, and are easily adsorbed onto suspended particulate matter (SPM). The deposition of SPM leads to the accumulation of considerable amounts of trisubstituted organotins and their degradation products in sediment (Hoch 2001). Several studies have been conducted on organotin pollution of river-, lake-, and harbor-sediments. Brack (2002) investigated organotin compounds in sediments from the Goteborg harbor, Sweden, and reported that their levels ranged from 17 to 366 ng/g dwt for TBT and from 1.5 to 71 ng/g dwt for TPT. These results were similar to those recorded from other harbors and marinas, and from an earlier study in the Goteborg harbor, which is located in the estuary (Brack 2002). DBT, MBT, DPT (diphenyltin), and MPT (monophenyltin), which are the degradation products of TBT and TPT, were also found in this harbor. TBT concentrations are the highest in the inner harbor and in the upper ~10-cm sediment layer. This indicates that there is a risk of TBT mobilization from the sediment surface, which may be exacerbated by the frequently disturbed harbor environment.

Takashi et al. (1997) studied the chemical speciation of organotin compounds that exist in sediments at a marina in Tokyo, Japan. These authors reported that >20 organotin compounds, including biodegraded ones, existed at the sampled site, and their identity was confirmed against authentic standards using gas chromatography/mass spectrometry (GC-MS) and a GC/atomic emission detection (GC-AED) system. Eleven organotin compounds were found in the Technical TBTChloride. Among them were unexpected organotin compounds, such as di-n-butyl (2-methylhexyl)tin chloride and di-n-butyloctyltin chloride.

The half-life of TBT in sediments is in the range of years. The accumulation of organotin on suspended particulates or sediments makes them available to filter- or sediment-feeding organisms. Resuspension of contaminated sediment offers an additional risk to aquatic organisms (Hoch 2001). The accumulation in sediments of butyltin and phenyltin species constitutes an ongoing pollution source, because residues of these compounds are slowly released into aquatic systems (Chiron et al. 2000; Ceulemans and Adams 1995; Kuballa et al. 1996).

3.3 Organotin in Organisms

Previous studies have revealed that high concentrations of toxic organotin compounds exist in some fish and aquatic invertebrates, such as gastropods and filter-feeding organisms. The presence of high concentrations of the toxic organotin residues in invertebrates results in imposex. Little is known about the accumulation and toxic effects of organotin in high trophic-level vertebrate predators; hence, their ability to disrupt endocrines of organisms worldwide is of concern. Humans are also exposed to the OTCs. The major route of such exposure is through food ingestion or exposure to household materials containing or contaminated by the organotins.

Hu et al. (2006) studied trophic magnification of TPT in a marine food web of Bohai Bay, North China; five benthic invertebrate species and six fish species were investigated. The concentrations of TPT detected in marine fish were, as expected, higher than those of TBT. A positive relationship was also found between trophic level and the concentration of TPT, indicating trophic magnification (TMF) of TPT in this food web.

Analysis of organotin residues in water and surface sediment samples from the bay revealed low environmental inputs of TPT, which indicated that the high concentrations of TPT found in fish from Bohai Bay resulted from food web magnification. The species in the study were primary producers (phytoplankton/seston and zooplankton) and comprised the following: five invertebrates: crab (*Portunus trituberculatus*), burrowing shrimp (*Upogebia* sp.), short-necked clam (*Ruditapes pluillippinarium*), veined rapa whelk (*Rapana venosa*), and bay scallop (*Argopecten irradians*). The other six species included the weever (*Lateolabrax japonicus*), catfish, (*Chateau - ichthys stigmatias*), bartail flathead (*Platycephalus indicus*), flower croakers (*Nibea albiflora*), wolfish (*Odontamblyopus rubicundus*), and mullet (*Lisa so-iuy*).

Zhang et al. (2003) worked on the butyltins in sediments and biota collected from the Pearl River Delta, South China. Both sediment and biota samples were collected and assessed using GC-AED analysis. The concentrations of TBT detected in the sediments ranged from 1.7 to 379.7 ng/g dwt. Shipping activities in the bay were thought to be responsible for the spatial distribution of the detected residues. A good linear relationship was observed between the reside ratios of DBT, TBT, and MBT samples taken from the Pearl River and associated estuary, and from the West River, suggesting a common source for the residues. All TBT concentrations in fish, mussel, and shrimp samples, which were collected in the study, retained residues that were below the seafood tolerable average residue level (TARL).

Meng-Pei et al. (2003) investigated the accumulation of OTCs in Pacific oysters (*Crassostrea gigas*), and both butyltin and phenyltin residues were quantified in this species. These oysters were collected during different seasons at several aquaculture sites, located along the west coast of Taiwan. Butyltin compounds were detected in oyster samples at all but one site. MPT and DPT compounds were not detected in any of the samples. The average concentration range of MBT, DBT, TBT, and tetrabutyltins (T$_4$BTs) in the sampled oysters was from non-detectable (n.d.) to 406 $\pm$ 12.7, n.d. to 280.9 $\pm$ 15.3, n.d to 417.2 $\pm$ 11.2, and n.d. to 85.8 $\pm$ 8.3 ng g^{-1}

(wwt), respectively. The concentration of TBT compounds detected in the oysters varied both spatially and temporally.

Lisicio et al. (2009) used two different analytical methods to determine levels of organotin compounds in marine organisms. Both methods involved extraction by tropolone, derivatization, and purification on Florisil™, followed by analysis using GC-MS. The main difference between the two procedures used was in the derivatization step: one employed a Grignard reagent (*n*-pentylmagnesium bromide), whereas the other method used sodium tetraethylborate (STEB). All compounds analyzed showed lower detection limits with STEB derivatization, particularly with TBT. Lisicio et al. (2009) also performed an in vivo experiment on TBT. He exposed one mussel species (*Mytilus galloprovincialis*) to known amounts of TBT for several days; both control and contaminated tissues were then analyzed using the STEB derivatization method. Results indicated bioaccumulation of TBT, which accumulated especially in the gills.

Albalat et al. (2002) assessed the levels of organotin pollution along the Polish coast (Baltic Sea), using mussels and fish as sentinel organisms. TBT, MBT, and DBT and TPT were the target compounds for which monitoring was performed. The bioaccumulation patterns found for the butyltin and phenyltin compounds varied substantially. The butyltins were detected in mussels at all sampled stations. Mussels sampled in the Gulf of Gdansk had the highest residue levels (68 ng/g wwt, measured as Sn) and had elevated TBT/DBT ratios, which suggested that there had been recent inputs of TBT to the area. Additionally, flatfish were sampled in the Gulf of Gdansk, and several tissues (liver, digestive tube, and gills) were individually analyzed. Although TPT residues were not detected in mussels in the Gulf of Gdansk, they were present in fish tissues. The highest organotin concentrations were observed in the liver (69 ng/g wwt, measured as Sn) of fish caught near the port at Gdansk. Relatively high concentrations were observed in the digestive tube, suggesting that organotin-contaminated food had been ingested, and food sources comprised an important uptake route of those compounds by mussels. Cooke (2002) studied the effect of organotins on human aromatase activity in vitro. TBT, at concentrations of 12 and 59 μM, and DBT, at a concentration of 74 μM, inhibited aromatase activity in vitro. In contrast, other organotins, such as MBT and the tri-, di-, and monooctyltins, were without effect.

3.4 Organotin in Soils

TPT acetate and TPT hydroxide have increasingly been used as soil treatment fungicides worldwide to treat a variety of crops. Such treatments have resulted in increasing levels of TPT acetate and TPT hydroxide in soils. Few studies have been conducted in which the abundance and persistence of TPT in soil has been measured. Kannan and Lee (1996) conducted a study on the foliage and soils of Pecan trees after application of TPT hydroxide. Their study results revealed that total phenyltin (MPT, DPT, and TPT) levels in foliage and soils ranged between

72 and 76 μg g^{-1} (Sn) dwt. In addition, TPT residues were reported in fish (blue gill, largemouth bass, and channel catfish) taken from a pond near a recently treated Pecan orchard (Visoottiviseth et al. 1995). The vapor loss during field spraying of TPT hydroxide is negligible because of its low vapor pressure (1×10^{-7} mm Hg at 25°C). But TPT is photolytically degraded in soils only if it is near the soil surface, where light can penetrate (Visoottiviseth et al. 1995).

3.5 Effects of Organotins in the Environment

The European Food Safety Authority (EFSA 2004) has assessed the health risk to consumers associated with exposure to the OTCs. It was concluded that the critical toxicological endpoint is immunotoxicity. Because different OTCs are similar to one another, they are grouped for risk assessment purposes. The tolerable daily intake (TDI) for the group was established as 250 ng/kg body weight and applied to the sum of residues that contain TBT, DBT, TPT, and di-n-octyltin (DOT). Alzieu (2000) reported that contact exposure to TBT causes irritation of the eyes and skin, potentially leading to severe dermatitis. Because of these properties, it is difficult to guarantee a safe environmental level for TBT. This means that its use as a biocide in aquatic systems could be incompatible with protecting ecosystems, preventing damage to certain marine activities, such as oyster farming.

Organotin compounds produce various known effects on aquatic organisms when they are exposed to these substances. These effects include larval mortality (Bella et al. 2005a, b), growth impairment, developmental and reproductive effects, and survival reduction in many marine species (Haggera et al. 2005). In addition, the results of animal experiments have suggested what the spectrum of potential adverse chronic effects of the organotins on humans may be. Among effects that could be damaging to humans are primary immunosuppression, endocrinopathy, neurotoxicity, metabolic effects, and effects on enzymatic activity. OTC exposure may also induce adverse effects on the eyes, the skin, the blood (dyscrasias), liver, and kidney, and on the following organ systems: cardiovascular, upper respiratory, gastrointestinal, and reproductive/developmental. Moreover, there is a risk of bioaccumulation and possibly carcinogenicity from OTC exposure (WHO-IPCS 1999; EU-SCOOP 2006; Nakanish 2007).

4 Fate of Organotins in the Environment

There have been several investigations into how the OTC compounds are distributed and degraded in the natural environment, and such information is both useful and important (Hoch 2001).

The OTCs enter ecosystems after marine or agricultural applications or after industrial use and release. However, research to date has focused only on tributyl- and triphenyl-tin pollution, because these compounds directly enter the environment through industrial use of organotin biocides. Recently, sewage sludge, municipal and industrial wastewater, and landfill leachates have also been discovered to constitute major sources of environmental organotins (Hoch 2001). Once these compounds become ecosystem pollutants, they may persist for long periods. How long they persist is a function of the status of various removal mechanisms. Removal mechanisms include physical ones (adsorption to suspended solids and sediments), chemical ones (i.e., chemical and photochemical degradation processes), and biological ones (i.e., uptake and biological degradation).

4.1 Degradation

The degradation of organotins in the environment occurs as a progressive elimination of organic groups from Sn cations. As successive organic groups are removed, toxicity is generally reduced. Degradation is achieved by both biotic and abiotic factors. Photodecomposition by ultraviolet (UV) light is the most important abiotic degradation process. In aquatic and terrestrial ecosystems, biological processes are the most important factor effecting degradation of the OTCs. Research has shown that organotin degradation is mediated by microorganisms; however, little information is available about the mechanism by which such degradation occurs. Also lacking is an understanding of the mechanism by which microbes are tolerant to the OTCs or the role played in degradation by anionic radicals (Dubey and Roy 2003). Biotic processes probably represent the most significant mechanisms by which TBT degradation occurs in soil, in freshwater, and in marine and estuarine environments (Dowson et al. 1993).

Research interest on the bioaccumulation and biodegradation of organotin in the water column, in sediments, and in marine organisms has been stimulated by the paucity of data available in these areas. Organotin compounds are known to be present in three main compartments of aquatic ecosystems: the surface microlayer, the water column, and at the surface layer of bottom sediments (Clark et al. 1988). TBT degrades rapidly to DBT and MBT, with half-lives of several days (Dubey and Roy 2003). The half-life value for the decline of TBT ($0.03\ \mu g^{-1}$) from a clean water site was 9 and 19 days for light and dark treatments, respectively (Dubey and Roy 2003). A first-order multistep kinetic model was used to describe the sequential degradation rate and pattern of TBT to form DBT, MBT, and tin (IV). Using this model, the half-lives of TBT, DBT, and MBT were 2.1, 1.9, and 1.1 years, respectively (Sarradin et al. 1995).

Abiotic degradation processes constitute other potential pathways for the degradation of TBT from soil, sediments, and water columns. Such abiotic processes may attack the Sn–C bonds by several different processes. Examples are UV irradiation-facilitated breakdown, chemical cleavage, gamma irradiation, and thermal cleavage. Only UV radiation (300–350 nm), in which the energy level corresponds to about $300\ kJ\ mol^{-1}$, is likely to cause direct photolysis of TBT. Because UV light does not

penetrate deeply, photolysis is expected to occur only in the upper few centimeters
of the water column (Clark et al. 1988). Maureen and Willingham (1996) reported
that the TBT degradation process may be explained as a sequential loss of an alkyl
groups from TBT to form toxic inorganic tin, as depicted immediately below:

$$R_3Sn^+ \rightarrow R_2Sn^{2+} \rightarrow RSn^{3+} \rightarrow Sn\,(IV)$$

TPT has low mobility, low solubility, and a strong ability to bind to soil and sediment in the aquatic environment (Blunden et al. 1986). For unbound organotins that
can be reached by chemical action, chemical cleavage may be mediated by mineral
acids, carboxylic acids, and alkali metals. These agents are capable of heterolyti-
cally cleaving Sn–C bonds, through both nucleophilic and electrophilic reactions
(Blunden and Evans 1990). Albalat et al. (2002) have studied the biodegradation
of the organotins. They monitored levels of TBT, MBT, and DBT at 10 stations
along the Polish coast (Baltic Sea). One mussel (*Mytilus edulis*) and one fish species
(*Platichthys flesus*) were used as sentinel organisms. The bioaccumulation patterns
of butyltin and phenyltin compounds varied substantially. Butyltin compounds were
detected in mussels from all sampled stations. TPT was not detected in mussel
but was found in fish, which indicated that ingesting organotin-contaminated food
was an important uptake route of those compounds in *P. flesus*. Paton et al. (2006)
investigated the microbial and chemical degradation and toxicity of phenyltin compounds in soil. These authors discovered that the degradation of organotins was
significantly slower in sterile soils vs. nonsterile soils. In nonsterilized soils, the
half-life of TPT was 27 and 33 days at amendment levels of 10 and 20 mg kg^{-1} Sn,
respectively. There was an increase in observed toxicity as the degradation of triphenyltin proceeded. This phenomenon proved that the metabolite formed is either
more bioavailable or more toxic than is the parent compound, or both.

4.2 Bioaccumulation

Lipophilicity is a criterion for the environmental persistence of organotins. Among
the organotins, TBT is considered to be an important pollutant because of its extreme
toxicity to several organisms and because of its tendency to bioaccumulate. Bacteria
have been reported to display a remarkable ability to accumulate TBT. Marine
bivalves are also able to accumulate significant amounts of TBT (up to 5 μg g^{-1}).
But fish and crustaceans accumulate much lower amounts, owing to their possession
of efficient enzymatic mechanisms to degrade TBT (Laughlin 1996). Absorption in
mice is also low, and TBT is mainly excreted unchanged via the feces. Mammals and
birds accumulate high levels of the butyltins in their organs and tissues (Iwata et al.
1995). In mammalian species, TBT compounds may be metabolized to DBT and
related metabolites. An undetermined amount of this compound is known to remain
in fat, liver, and kidney (Adeeko et al. 2003). Other researchers have undertaken
studies to evaluate the bioaccumulation of organotins (Harino et al. 2005; Strand

et al. 2005; Azumi et al. 2007). Similar results to those of Adeeko et al. (2003) were recorded in these other studies.

4.3 Sorption of the Organotins and Their Biological Effects

In recent years, restrictions have been placed on the use of TBT on pleasure boats in Europe. Although considerable progress has been made in reducing TBT effects, they still continue to be observed in marine ecosystems. An essential source of contamination of TBT along the German North Sea and the Baltic Coast has been remobilization (by desorption) of the high TBT concentrations present in sediments (Langston and Popoe 1995).

A comparison of the burden of TBT in sediments to which snails and mussels are exposed gives rise to concern for conducting any future dredging and disposal of TBT-contaminated sediments (WWF 1995). Because suspended matter has a high affinity for organotin compounds, any perturbation of sediments by dredging may remobilize TBT and thereby substantively increase TBT residue levels in the water column. Presently, desorbed or actively remobilized TBT-contaminated sediment, in harbors and in some coastal areas, constitutes the main source of biologically available TBT (Langston and Popoe 1995).

Hongwen et al. (1996) investigated the adsorption behavior of eight organotin species and Sn^{4+} ($SnCl_4$) on estuarine sediments. They found that adsorption of the organotins varies greatly and depends on molecular structure. The order of adsorption coefficient for tin compounds in the studied sediment samples was as follows: tetra $\rightarrow$ mono $\rightarrow$ di $\rightarrow$ triorganotins. Correlations of the log K values (using eight different structural parameters) showed that the electronic properties of the Sn atom constitute the principal factor controlling their adsorption behavior.

The mechanism by which the organotins are adsorbed is mainly through an ion exchange process and involves little lipophilic partitioning (Hongwen et al. 1996). Hermosin et al. (1993) reported the adsorption mechanisms of MBT to various clay minerals and found that its adsorption capacity for all clays was higher than the corresponding cation exchange capacity (CEC value).

Adsorption onto clay is important to the environmental distribution and fate of organotins, because research has shown that large proportions of organotin contaminants are associated with the clay fraction of particulate matter. Thus, soils and sediments may serve as traps for these toxic contaminants. Unfortunately, the number of studies conducted on the remobilization of adsorbed organotin from environmental media is still few (Hoch 2001).

4.4 Biomethylation

Methyltin compounds can be formed by processes that involve biomethylation. Several biotic and abiotic methylation agents exist. Methylcobalamin (CH_3B_{12}), the methyl co-enzyme of vitamin B_{12}, is a carbanion donor that is able to convert inorganic Sn (IV) to several methyltin species (Hoch 2001). Methylcobalamin has been demethylated by $SnCl_2$ in aqueous HCl solution, in the presence of an oxidizing agent (Fe^{3+} or Co^{3+}), to form a monomethyltin species. Methyliodide (CH_3I) can also methylate tin species, whereas tin (IV) compounds do not so react. Chemical or biological processes are capable of methylating inorganic tin (II), Sn (IV), and methyltin derivatives under stimulated environmental conditions. Recently, methylation of butyltin species in sediments has been reported (Hoch 2001) and may arise from biological methylation of anthropogenic butyltins in the aquatic environment. Selected possible reactions of Sn–C include the following:

$$2Bu_3Sn^+ \rightarrow Bu_2Sn^{2+} + Bu_4Sn \tag{1}$$

$$Bu_2Sn^{2+} + Bu_3Sn^+ \rightarrow BuSn^{3+} + Bu_4Sn \tag{2}$$

$$Bu_3Sn + Me^- \rightarrow Bu_3MeSn \tag{3}$$

Biomethylation processes are of great ecological relevance, because some methylated metals have higher toxicity to aquatic organisms than does the inorganic metal (Hoch 2001).

5 Fate of Organotins in Marine Invertebrates

5.1 Bioaccumulation in Marine Invertebrates

Most research on TBT accumulation by marine invertebrates was concentrated on mollusks (bivalves) and crustaceans (decapods), because these groups dominate the ecological habitat and serve as important seafood resources (Laughlin 1996). Research conducted on TBT accumulation by marine invertebrates revealed that marine bivalves are able to accumulate significant amounts of TBT (up to $> 5 \ \mu g \ g^{-1}$) (Laughlin 1996). Azumi et al. (2007) studied the accumulation of organotin compounds at aquaculture sites in Korea. High concentrations of butyltin compounds (mono-, di-, and tri-butyltins) were detected, especially in the gills, hepatopancreas, and digestive tracts of sea squirt (*Halocynthia roretzi*).

Meng-Pei and Shin-Mei (2003) investigated levels of OTCs in Pacific oysters (*Crassostrea gigas*) collected from aquaculture sites. Butyltin compounds were detected in most samples, whereas no MPT and DPT compounds were detected. The average concentrations of monobutyl-, tributyl-, triphenyl-, and tetraphenyltins ranged from detectable (n.d.) to 406.6 ± 12.7, n.d. to 28.09 ± 15.3, n.d. to 417.2 ± 11.2, and n.d. to $85.8 \pm 8.3 \ ng \ g^{-1}$ (wwt), respectively.

The accumulation of OTCs also occurred in deep-sea organisms, namely gastropods (*Colliloconcha nankaiensis*), sea cucumbers (*Psychropotes verrucosa*), galatheid crabs (*Munidopsis albatrossae* and *Munidopsis subsquamosa*), *and* bivalves (*Calyptogena tsubasa* and *Calyptogena nautilei*). High concentrations of BT and PT (phenyltin) compounds were observed in gastropods and sea cucumbers. The composition of BT in deep-sea organisms was calculated, and an increase in the MBT proportion was recorded, while a decline in DBT proportion was observed at higher trophic levels (Harino et al. 2005). Accumulation of organotins in marine invertebrates has also been reported by Harino et al. (2008). The concentration of OTC in seven species of dolphin (bottlenose, finless porpoise, Indo-Pacific hump-backed, long-backed common, Pantropical spotted, spinner and striped), which were stranded on the coast of Thailand, was measured. The ratio of the average of BT and PT compounds in tissues and organs was 16:1; average residue levels in tissues and organs for the dolphins were 152 and 62 μg kg^{-1}, respectively. The highest concentration of TBT was generally observed in the liver. No significant difference in the concentration of OTC between genders was observed. The concentrations of BTs in all organisms were high and of following order: whales > dugongs > dolphins. The concentrations of PTs in whales were higher than those in dolphins and dugongs. In general, it has been observed that species with a high rate of uptake or a low rate of metabolic conversion and elimination display relatively high bioaccumulation ratios (Meador and Rice 2001).

5.2 Toxicity to Marine Invertebrates

TBT causes impairments in growth and development, and induces reproductive failures, shell anomalies, and gel formation. It also causes chambering and high mortality, disturbs the energy metabolism of bivalves, and inhibits the activity of many enzymes; these effects reduce the survival of many species (Beaumont and Budd 1984; Haggera et al. 2005). TBT, as early as the 1970s, was known to be very toxic to many aquatic organisms (Blabber 1970; Smith 1981). The high toxicity of TBT is attributed to its effects on mitochondrial function (Blabber 1970; Smith 1981). The embryonic and larval stages of marine invertebrates are less tolerant to toxicants than are adults, and this difference has been used to assess the biological quality of marine water and sediments (Fent and Muller 1991).

TBT is known to have other toxic endpoints (Horiguchi et al. 1998), e.g., acute lethal toxicity in rock shell larvae (*Thais clavigera*). However, growth impairment is a much more sensitive endpoint for measuring exposure to TBT than is mortality (Meador and Rice 2001). TBT is known to inhibit oxidative phosphorylation, which affects cell metabolism by stimulating the production of adenosine diphosphate, and results in mitochondrial membrane malformation.

TBT affected larval development of bivalves (*C. tsubasa*) and caused sexual disturbances in gastropods (*C. nankaiensis*) at nanogram per liter levels in seawater. At a level of 1.0 ng L^{-1}, TBT caused masculinization in many female gastropods (*C. nankaiensis*), a phenomenon known as imposex. It also limits cell division in phytoplankton and reproduction of zooplankton. TBT has been reported to induce shell calcification anomalies in the oyster *Crassostrea gigs* at a level of 2 ng L^{-1} and to disturb the reproduction of bivalve mollusks at 20 ng L^{-1} (Bella et al. 2005a). Ruiz et al. (1995) investigated the effect of TBT exposure on veliger larval development of the bivalve (*C. tsubasa*). They found that TBT contributed to the demise of clam populations by preventing successful and timely development of veliger larvae. TBT also affects the abundance and relative growth rates of male and female whelks around marinas (Gil et al. 2000).

6 The Role of Biomarkers

Pollution of the marine environment is a global concern because of the adverse effects caused by various contaminants, whose levels are growing at an alarming rate. Residues of many contaminants, such as the OTCs, continue to enter the natural environment and continue to accumulate in many organisms. Therefore, it is crucial that means to track both the presence and effects of such contaminants be developed. Biomarkers offer one important way in which environmental contaminate effects can be monitored.

The idea behind biomarkers is not a new concept but is a new name for a preexisting monitoring principle (Adams 1990). Biomarkers are defined as the measurements of body fluids, cells, or tissues that indicate, in biochemical or cellular terms, the presence of contaminants or the magnitude of the host response (Bodin et al. 2004). According to Van Gestel and Van Brummelen (1996), "biomarkers" are any biological response to an environmental chemical that is measured inside an organism or its products (urine, feces, hairs, feathers, etc.), and indicates a departure from the normal status. A response may result from a biochemical, a physiological, a histological, and/or a morphological (including appearance, pigmentation, and surface deformation) measurement of health, although behavioral effects are excluded. Hence, biomarkers cannot be used to measure effects in intact organisms or cause affected organisms to deviate from their normal status (Van Gestel and Van Brummelen 1996). Therefore, one can discern that biomarkers are potentially sensitive tools of immense importance for measuring biological effects that affect environmental quality (Sarkar et al. 2006).

Some authors claim that biomarkers may also be accommodated into whole animal studies (Ross et al. 2002; Magni et al. 2006) and may be specific to one pollutant or may be altered in response to either pollutant effects or the presence of natural stressors (Pfeifer et al. 2005). What is certain is that they are potentially very useful as prognostic and diagnostic early warning tests and offer the potential of specificity, sensitivity, and application to a wide range of organisms (Sarkar et al. 2006). The use of properly researched biomarkers is not limited to laboratory use but may be applied to field studies too. However, the initial development of biomarkers usually involves laboratory experimentation to first identify potential responses, and to establish causal mechanisms, before application to field use (Sarkar et al. 2006).

6.1 The Significance and Utility of Biomarkers

Biomarkers are used to evaluate the exposure effects of many different contaminants (i.e., metals, organic xenobiotics, and organometallic compounds) (Ross et al. 2002; Depledge and Fossil 1994). The most significant features of the use of biomarkers are summarized below:

1. They offer means to achieve sensitive detection of selected chemical stresses within organisms.
2. They generate insights on possible harmful effects that cannot be obtained from chemical analysis alone (Depledge and Fossil 1994).
3. They may be used to predict effects on invertebrate populations and communities (Largardic et al. 1994), and may help assess types or degree of environmental damage, or formulation of regulations to control such damage (Sarkar et al. 2006).
4. They offer means to identify interactions between contaminants and organisms, and measure sublethal effects (Sarkar et al. 2006).
5. They offer alternative ways of detecting the presence of both known and unknown contaminants (Sarkar et al. 2006).
6. They constitute a temporally and spatially integrated measure of the degree to which pollutants are bioavailable (Sarkar et al. 2006).
7. They may be used to establish important routes of exposure by application to species from different trophic levels and aid in designing strategies for intervention and remediation (Sarkar et al. 2006).

6.2 Biomarkers of TBT in Marine Invertebrates

Useful biomarkers have been developed to help monitor the effects of contaminants in marine invertebrates. Among these are the following biomarkers that have been used to assess the toxicity of TBT: metallothionein induction, acetyl cholinesterase inhibition, imposex, lysosomal enlargement, lysosomal membrane destabilization, peroxisome proliferation, lysosomal activity, genetic or molecular biomarkers, TBT-sensitive immunological biomarkers, apoptosis induction, phagocytic index, and amoebocytic index.

Some of these biomarkers are more useful than others. Below, we provide greater detail on prominent types of these.

6.2.1 Metallothionein (MT) Induction

MTs are cysteine-rich peptides that exist in the cytosol and the nucleus and in lysosomes. They are non-enzymatic proteins that have low molecular weight, no aromatic amino acids, and are heat stable (Olsson et al. 1998; Roeva et al. 1999). MT-like proteins have been reported in many aquatic invertebrates but occur mainly in mollusks (Isani et al. 2000). Mussels, used worldwide in environmental pollution assessment, are good candidates for monitoring MT for assessment of metal contamination (Leinio and Lehtonen 2005; Raspor et al. 2004; Mourgand et al. 2002; Petrovic et al. 2001). The use of MT as a biomarker has been validated in many in situ studies (Lionetto et al. 2001; Petrovic et al. 2001; Rodriguez-Ortega et al. 2002; Ross et al. 2002; Mourgand et al. 2002). Such studies have generally found MT to work well for the purpose intended. Fafandel et al. (2003) investigated molecular response to TBT stress in marine sponges (*Suberites domuncula*). Proteolytic cleavage and phosphorylation of stress response KRS–SD protein kinase in control and TBT-treated sponges were investigated. Exposure of sponges to TBT resulted in alteration of KRS–SD1 and KRS–SD2 expression levels and their phosphorylation state. KRS–SD induction, its phosphorylation, and proteolytic cleavage during TBT stress suggest that in sponge cells, mechanisms exist similar to ones present in human cells in which KRS/MST protein kinase is involved in promotion of apoptosis following oxidative stress.

6.2.2 Acetyl Cholinesterase (AChE) Inhibition

AChE enzymes are responsible for hydrolyzing the neurotransmitter acetylcholine into choline and acetic acid. AChE is usually located in the membranes of erythrocytes of both vertebrates and invertebrates. AChE controls ionic current in excitable membranes and plays an essential role in nerve condition at the neuromuscular junction (Pfeifer et al. 2005; Magni et al. 2006). AChE biomarkers may be less useful in fish, because fish have higher levels of tolerance to AChE inhibition. Measurements of AChE inhibition are most frequently used where a biomarker for organophosphate insecticide exposure is required (Matozzo et al. 2005).

However, AChE biomarkers have also been used with the OTCs. Rebeiro et al. (2002) evaluated TBT subchronic effects in tropical freshwater fish (Linnaeus *Astyanax bimaculatus*). Linnaeus *A. bimaculatus* adult fish were acclimatized in a laboratory and isolated into groups of eight individuals. Two groups were used as controls and one group was exposed to TBT chloride, dissolved in corn oil (0.0688 $\pm$ 0.0031 μg TBT g^{-1}), every 6 days for 32 days. A muscle fragment was excised for the determination of the acetylcholinesterase activity and blood smears were obtained for differential while cell counts. The results indicated nuclear irregular shapes, chromatin condensation, presence of intranuclear lipid bodies, and degenerative nuclei. AChE activity was not affected by TBT exposure. The increasing number of metrophilis may represent cytotoxic and stress conditions facilitating the invasion of the opportunist.

6.2.3 TBT-Sensitive Immunological Biomarkers

Several xenobiotics alter immune function and the immune system. TBT has been observed to have adverse effects on cellular immune functions of hemocytes. The three indices established as TBT pollution biomarkers are amoebocytic index, phagocytic index, and lysosomal activity index (Chima et al. 1999).

6.2.4 Lysosomal Biomarkers

Matozzo et al. (2002) studied the effects of TBT on circulating cells from the clam *Tapes philippinarum*. They found that exposure of hemocytes to 0.05 μm TBT caused a significant increase ($p < 0.05$) in neutral red dye uptake into the lysosomes, compared with controls, whereas exposure to TBT caused no differences. Enlarged lysosomes were observed in hemocytes exposed to TBT. Moreover, in hemocytes treated with 0.05 μm and 0.1 μm of TBT, superoxide chromatase activity significantly decreased ($p < 0.05$ and $p < 0.1$, with respect to that of the control. A significant decrease in lysozyme activity was also observed in hemocytes exposed to 0.05 and 0.1 μm TBT. Lysozyme is a lysosomal enzyme that may be secreted by hemocytes in the hemolymph during phagocytosis. Reduced lysozyme activity suggests immunosuppression, resulting in lowered resistant bacteria challenge (Matozzo et al. 2002).

6.2.5 Molecular (genetic) Biomarkers

Because pollutants interact with the receptors of organisms at the molecular level to cause their effects, the measurement of certain molecular biomarkers may have obvious advantages for detecting early chemical effects (Nicholson and Lam 2005).

Schroth et al. (2005) utilized a strategy that identified molecular biomarkers and linked the study of abiotic stress to evolutionary history. These authors used the Moon jellyfish, *Aurelia* spp., as a model species. The authors used complementary DNA subtraction analysis to identify genes that were differentially regulated after exposure to the chemical stressor TBT. They also identified differential expression patterns following exposure to TBT at different temperatures. Results suggested that the identified genes were involved in response to the chemical, as well as to heat-induced stress.

6.2.6 Apoptosis

This is a form of genetically programmed cell death, which can be initiated by an internal clock or by exposure to extracellular agents such as hormones, cytokines, killer cells, and a variety of chemical and viral agents. These methods that are applied when using apoptosis as a biomarker are normally characterized by morphological and biochemical criteria (Micic et al. 2001).

Micic et al. (2001) investigated the induction of apoptosis by tri-nTBT in gill tissue of the mussel *M. galloprovincialis*. These authors used the terminal dUTP nick-labeling technology (TUNEL) to detect cells displaying DNA fragmentation within gill structures. Genomic DNA fragmentation was detected as characteristically ladder-like patterns of DNA fragments that were induced by a single injection directly into the pallial fluid of different doses of TBT below the mantle, after 1 day of incubation.

After 1.5 h of TBT incubation, DNA degradation of a higher order DNA structure and a reduced G_0/G_1 cell cycle region were detected. The effect of TBT on the cell cycle in the mussel (*M. galloprovincialis*) gill was dose related and exposure time dependant. In this study, three types of investigation were performed: (a) detection of internucleosomal fragmentation by conventional gel electrophoresis, (b) identification of DNA fragments of higher chromatin organization by pulsed-field gel electrophoresis, and (c) the detection of apurinic sites in gill sections of TBT-treated mussel (*M. galloprovincialis*) using TUNEL. The process of apoptosis in vivo induction in the blue mussels (*Mytilus galloprovincialis*) was described for the first time (Micic et al. 2001).

6.2.7 Imposex

Imposex is characterized by the development of morphological features (i.e., penis and vas deferens) in female gastropod mollusks or superimposition of male morphological features onto females. Imposex results from exposure of certain invertebrates to organotin antifouling paints (Marshall and Rajkumar 2003). Imposex serves as a useful morphological biomarker for measuring organotin contamination of marine ecosystems. High incidences of imposex were characterized by lower female to male ratios, suggesting that sterility and female mortality were TBT related (Marshall and Rajkumar 2003). In other studies, organotins were found to accumulate in the tissue of marine invertebrates. TBT generally shows the greatest accumulation among the butyltin compounds and is the primary cause of imposex (Bryan et al. 1988; Barreiro et al. 2001). The induction of imposex by TBT may account for a sizable portion of the decline of certain coastal marine mollusks (Gibbs and Bryan 1996).

Pessoa et al. (2001) studied the occurrence of organotin compounds in Portuguese coastal waters and found that acute effects from TBT were induced at concentrations as low as 1 μg/L in aquatic organisms; moreover, imposex was induced at levels below 0.5 ng/L of TBT (as Sn). TBT at 20 ng/L (as Sn) caused sterility, and this was followed by the disappearance of the most sensitive neogastropods on a given shore. The authors concluded that the use of imposex was the most sensitive indicator of exposure to TBT of all known non-target pathological conditions.

7 The Regulation of Organotin Compounds

The presence of tributyltin in the environment has attracted the most regulatory attention because of the volume of its use in antifouling paints to coat boat hulls or harbor edifices. When biocides are released from paint over time, it forms a thin layer of concentrated TBT in the vicinity of its immediate use area. This contaminated area repels or kills organisms such as barnacles (Huggett et al. 1992). Moreover, TBT diffuses from the application area to contaminate adjacent water, sediments, and non-target organisms. As previously mentioned, TBT contamination causes morphological aberrations in oysters and mussels (Wadlock and Thain

1983). These effects and other associated environmental impacts of TBT had led the authorities of many countries to target TBT for regulation (Abbott et al. 2000).

According to the USEPA (United States Environmental Protection Agency) (2001), TBT restrictions apply in many countries around the world. For example, the European Union, Canada, Scandinavia, and South Africa have banned the use of TBT on vessels that are less than 25 m in length. As a result of increasing awareness of the undesired effects of TBT, global efforts have been made to solve this problem, and increasingly, legal requirements have been enforced to protect the aquatic environment from TBT (Konstantinon and Albanis 2004).

France, in 1982, was the first country to ban the use of organotin in antifouling paints for application to boats of less than 25 m in length (Alzieu et al. 1986). This ban was sequel to the collapse of the oyster industry in France' Archon Bay in the late 1970s and early 1980s (Alizieu et al. 1989, 1991). The enhanced TBT concentrations in seawater and the frequency of oyster shell anomalies were the cause of the collapse. Subsequently, comparable regulations as those imposed in France were also passed, after 1988, in North America, UK, Australia, New Zealand, Hong Kong, and most European countries (Alzieu et al. 1989; Champ 2000, 2003; De Mora et al. 1995).

The International Maritime Organization (IMO) campaigned for a global treaty to ban the application of TBT-based paints starting 1 January 2003; as a result, a total prohibition took place by January 2008 (IMO 2001). In Europe, the current Water Framework Directive is the major community instrument for controlling port and diffused discharges of dangerous substances. Decision no. 2455/ 2001/EC (20 November 2001) of the European Commission Parliament amended the water policy directive 2000/ 60/EC and defined 11 priority hazardous substances, including TBT compounds, that were subject to cessation of emission, discharge, and lose to water.

In addition, regulation No. 782 /2003 of the European Parliament and of the council of 14 April 2003 was aimed at prohibiting organotin compounds on all ships entering European seaports. TBT monitoring was also mandated by legislation from several European Commissions, including the council decisions 75/437/EC (marine pollution from land-based sources), 77/585/EC (Mediterranean Sea), and 77/586/EC (River Rhine), and the council directive 80/68 EC (groundwater) (Champ 2000).

In 1985, the government of the United Kingdom (UK) prohibited the application of TBT-based antifouling paints to small vessels. In 1986, an Environmental Quality Target Concentration (EQTC) was set for TBT at a level of 20 ng L^{-1}. This value was based on the lethal concentrations that were effective for control of selected commercially important mollusks. Because of the high toxicity value of the TBT, this value was reduced by a factor of ten 1 year later to achieve improved environmental protection (Takahashi et al. 1997). In Spain, a Royal Decree (995/2000) established the concentration limit of organotin species in waste discharges to continental surface waters. The value selected was less than 20 ng L^{-1}. Legislation that addresses concentrations in seawater samples has yet to be approved.

The United States enacted the Organotin Antifouling Paint Control Act in 1988; a leaching rate of organotins from the application sites was limited to $4 \mu g \, cm^{-2} \, d^{-1}$ (USA 1988). Moreover, the Occupational Safety and Health Administration (OSHA), the American Federal Agency, and the National Institute for Occupational Safety and Health (NIOSH) have established workplace exposure limits of 0.1 mg m^{-3}. The Food and Drug Administration (FDA) has also set a limit for the use of tin as a food additive (ATSDR 2005). In addition, the water quality criterion of the USEPA is that aquatic life and the uses to which aquatic life are put should not be unacceptably affected.

In 1989, the Canadian government regulated TBT (under the Canadian Pest Control Products Act) by stipulating a maximum daily release rate for antifouling paints of 4 µg TBT per cm^3 of boat–ship hull surface. In Australia, the evidence for establishment of a relationship between deformities in oysters and the presence of TBT in oyster tissue led to the banning of TBT-based paints (Takahashi et al. 1997). Japan also restricted TBT usage on antifouling coatings of boats and aquaculture nets by implementing limits in 1990. But TBT is still used as an antifouling agent for ocean liners and deep-sea fishing boats (Takahashi et al. 1997).

Similar actions on the usage of TBT in paints were taken by Switzerland, the Netherlands, Sweden, New Zealand, South Africa, and most European countries (Sergi et al. 2005). However, the legislative restrictions on the use of TBT-based marine paints in Tanzania are less clearly defined. As a result of legislation restricting the use of TBT-based antifouling paints, some reduction in the levels of TBT has been reported, particularly in areas proximate to recreational shipping activities (Rees et al. 2001; Hawkins et al. 2000). However, in areas near industrial shipping activities (e.g., ports), TBT levels remain high (Valkirs et al. 2003; Peachery 2003; Horiguchi et al. 2004; Harino et al. 2006).

South Africa is positioned along a primary shipping route between Europe, the Americas, and Asia. South African harbors provide infrastructural support to the global shipping industry, with some of the largest and busiest African harbors being located on the eastern seaboard of South Africa. The Constitution of the Republic of South Africa (Act 108 of 1996) and the Bill of Rights enshrine basic human rights, such as having access to sufficient water and a safe and healthy environment. The two Acts that enable the South African government to fulfill these rights (through the Department of Water affairs) are the Water Services Act (Act 108 of 1997) and the National Water Act (1998). In South Africa, the Maritime International organization (IMO) held an international convention on the control of harmful antifouling systems in 1990. The convention was adopted in 2001, and South Africa was a signatory. The convention required prohibition or restriction of the application of antifouling systems and they listed the substances to be controlled. The convention also required signatory states to ensure that controlled substance application or removal was done appropriately and required such states to perform surveys of their own ships. The regulations required that any ships in violation of the convention standard were subject to being warned, detained, dismissed, or excluded from a country's port (IMO 2001).

This convention required the South African government to develop new legislation to effect provisions of the convention. The Annex 1 of the convention included a list of organotin compounds. The waste resulting from the removal of these toxins, as stated in Article 5 of the convention, should be disposed of in accordance with permits from the Department of Water affairs (DWA) and Environmental Affairs (DEA). The South African Maritime Safety Authority (SAMSA) became responsible for enforcing and implementing the legislation; the provision of waste disposal was taken over by the National Port Authority (NPA) (IMO 2001).

The Facilitation of International Maritime Traffic (FAL) 1991 amendments to the convention were passed to prevent unnecessary delays in maritime traffic. This required the port authority to inspect foreign ships to verify that their condition, manning, and operation were in compliance with international rules and the regulating act of the South African Maritime Authority. Several other conventions for protection of coastal and marine ecosystems are in force, and are indirectly related to organotin contamination. For example, the Ballast Water Convention requires that pollution checks be made of the maritime environment resulting from discharges of oil and other hazardous waste generated outside Africa into African countries. The Lome IV Convention also bans the export of hazardous waste from European countries to Africa (EC report 2007).

In general, despite the ban on, or regulation of, TBT usage in some countries, TBT contamination continues in the aquatic environment; therefore, environmental concerns for this contaminant remain high and warrant continued assessment and monitoring. Continued diligence is needed, particularly in countries that do not restrict the use of TBT-containing antifouling paints; moreover, further research is necessary on elucidating the pathways, kinetics, and persistence of organotin compounds.

8 Conclusions

In this chapter, we have reviewed the fate and distribution of, and the human exposure to, organotin compounds in the environment. The organotins, some of which are very toxic, have been confirmed as predominant pollutants of freshwater and marine ecosystems. The presence of these organotin residues in the environment is clearly undesirable. Researchers have defined the toxicity of many organotin compounds and have reported organotin residue to exist in both aquatic and terrestrial ecosystems. Although a considerable amount of research has been conducted on the response of marine species to organotins in water, only limited data are available on the deposition of butyltin in humans. This is disturbing, because there is evidence of human exposure to OTCs. Therefore, we conclude that additional research is needed in the following areas:

- the absorption kinetics in humans, mechanisms of action, and human exposure levels, along with body burdens of the organotins;
- additional studies of the toxicity of organotin compounds in water;
- investigations designed to better understand the effects of sediments on organotin exposure in aquatic organisms;

- further definition of the use of biomarkers that can delineate organotin toxicity in mussels;
- studies to define levels of organotin compounds that exist in foodstuffs;
- studies to better define the toxic responses of marine species to TBT residues; and
- an evaluation of the extent to which human exposure exist to organotins in the atmospheric environment.

9 Summary

Organotin compounds result from the addition of organic moieties to inorganic tin. Thus, one or more tin–carbon bonds exist in each organotin molecule. The organotin compounds are ubiquitous in the environment. Organotin compounds have many uses, including those as fungicides and stabilizers in plastics, among others in industry. The widespread use of organotins as antifouling agents in boat paints has resulted in pollution of freshwater and marine ecosystems. The presence of organotin compounds in freshwater and marine ecosystems is now understood to be a threat, because of the amounts found in water and the toxicity of some organotin compounds to aquatic organisms, and perhaps to humans as well. Organotin compounds are regarded by many to be global pollutants of a stature similar to biphenyl, mercury, and the polychlorinated dibenzodioxins. This stature results from the high toxicity, persistence, bioaccumulation, and endocrine disruptive features of even very low levels of selected organotin compounds.

Efforts by selected governmental agencies and others have been undertaken to find a global solution to organotin pollution. France was the first country to ban the use of the organotins in 1980. This occurred before the international maritime organization (IMO) called for a global treaty to ban the application of tributyltin (TBT)-based paints. In this chapter, we review the organotin compounds with emphasis on the human exposure, fate, and distribution of them in the environment. The widespread use of the organotins and their high stability have led to contamination of some aquatic ecosystems. As a result, residues of the organotins may reach humans via food consumption. Notwithstanding the risk of human exposure, only limited data are available on the levels at which the organotins exist in foodstuffs consumed by humans. Moreover, the response of marine species to the organotins, such as TBT, has not been thoroughly investigated. Therefore, more data on the organotins and the consequences of exposure to them are needed. In particular, we believe the following areas need attention: expanded toxicity testing in aquatic species, human exposure, human body burdens, and the research to identify biomarkers for testing the toxicity of the organotins to marine invertebrates.

Acknowledgments The authors wish to thank the management of Cape Peninsula University of Technology, Cape Town, South Africa for financial support. H.K. Okoro also acknowledges University of Ilorin, Ilorin, Nigeria, for supplementation staff development award.

References

Abbott A, Abel PD, Arnold DW, Milne A (2000) Cost benefits analysis of the use of TBT: the case for a treatment approach. Sci Total Environ 258:5–19

Adams SM (1990) Status and use of biological indicators for evaluating the effects of stress on fish. Am Fish Soc Symp 81:1–8

Adeeko A, Li D, Forsyth DS, Casey V, Cooke GM, Barthelemy J, Cyr DG, Trasler JM, Robaire B, Hales BF (2003) Effects of in utero tributyltins chloride exposure in the rat on pregnancy outcome. Toxicol Sci 74:407–415

Albalat A, Potrykins J, PempoKwiak J, Porte C (2002) Assessments of organotin pollution along the polish coast (Baltic Sea) by using mussels and fish as sentinel organisms. Chemosphere 47:165–171

Alzieu C (2000) Impact of tributyltin on marine invertebrates. Ecotoxicology 9:71–76

Alzieu C, Michel P, Tolosa I, Bacci E, Mec LD, Readman JW (1991) Organotin compounds in the Mediterranean: a continuing cause of concern. Mar Environ Res 32:261–270

Alzieu C, Sanjan J, Deltreil JP, Bonel M (1986) Tin contaminations in Archachon Bay- effects on oyster shell anomalies. Mar Pollut Bull 17:494–498

Alzieu C, Sanjuan J, Michel P, Borel M, Dieno JP (1989) Monitoring and assessment of butyltins in Atlantic coastal waters. Mar Pollut Bull 208:22–26

ATSDR (2005) Agency for toxic substances and diseases registry. Toxicological profile for tin and tin compounds. US Department of Health and Human Services. http://www.atsdr.cdc.gov/ToxProfiles/tp55-Cl.pdf.

Azuela M, Vasconcelos MT (2002) Butyltins compounds in Portuguese wines. J Agric Food Chem 50:2713–2716

Azumi K, Nakamura S, Kitamura S, Jung S, Kanehira K, Iwata H, Tanabe S, Suzuki S (2007) Accumulation of organotin compounds and marine birnavirus detection in korean ascidians. Fish Sci 73:263–269

Barreiro R, Gonzalez R, Quintela M, Ruiz M (2001) Imposex organotin bioaccumulation and sterility of female *Nassarius reticulatus* in polluted areas of NW Spain. Mar Ecol Prog Ser 218:203–212

Beaumont AR, Budd MD (1984) High mortality of the larvae of the common mussel at low concentrations of tributyltin. Mar Pollut Bull 15:402–405

Becker G, Janak K, Colmsjo A, Ostman C (1997) Speciation of organotin compounds released from poly(vinylchloride) at increased temperatures by gas chromatography with atomic emission detection. J Chromatogr A 775:295–306

Bella J, Beiras R, Marino-Balsa JC, Fernandez N (2005a) Toxicity of organic compounds to marine invertebrate embryos and larvae: a comparison between the sea urchin embryogenesis bioassay and alternative test species. Ecotoxicology 14:337–353

Bella J, Halverson A, Granmo A (2005b) Sublethal effects of a new antifouling candidate in lumpfish (*Cyclopterus lumpus* L.) and Atlantic cod (*Gadus morhua*) larvae. Biofouling 21:207–216

Blabber SJM (1970) The occurrence of a penis-like outgrowth behind the right tentacle in spent females of nucella lapillus. The Malacological Society of London, London

Blanca AL (2008) Environmental levels toxicity and human exposure to TBT- contaminated environment. A Rev Environ Int 34:292–308

Blunden SJ, Evans CJ (1990) Organotin compounds. In: Hitzinger O (ed) The hand- book of environmental chemistry, vol 3, Part E. Anthropogenic compounds. Springer, Berlin, pp 1–44

Blunden SJ, Hobbs LA, Smith PJ, Craig PJ (eds) (1986) Organometallic compound in environment. Longman, Harlow, p 111

Bodin N, Burgeot T, Stanisiere JY, Bolquene G, Menard D, Minier C, Boutet I, Amat A, Chard Y, Budzinski H (2004) Seasonal variations of a battery of biomarkers and physiological indices for the mussel *Mytilus galloprovincialis* transplanted into the northwest Mediterranean Sea. Comp Biochem Physiol Toxicol: Pharmacol 138(4):411–427

Brack K (2002) Organotin compounds in sediments from the Gota alv estuary. Water Air Soil Pollut 135:131–140

Bryan GW, Gibbs PE, Burt GR (1988) A comparison of the effectiveness of tri-n-butyltin chloride and five other organotin compounds in promoting the development of imposex in the dog whelk, I. J Mar Biol Assoc UK 68:733–744

Ceulemans M, Adams FC (1995) Evaluation of sample preparation methods for organotin speciation analysis in sediments- focus on MBT extraction. Anal Chim Acta 317: 161–170

Champ MA (2000) A review of organotin regulatory strategies pending actions, related costs and benefits. Sci Total Environment 258:21–71

Champ MA (2003) Economic and environmental impacts on ports and harbours from the convention to ban harmful marine antifouling systems. Mar Pollut Bull 46:935–940

Chieu LC, Hung TC, Choang KY, Yeh CY, Meng PJ, Sheih MJ, Hem BC (2002) Daily intake of TBT, Cu, Zn, Cd and as for fishermen in Taiwan. Sci Total Environ 285:177–185

Chima F, Marin MG, Mattozzo V, Ros DL, Ballarin L (1999) Biomarkers for TBT immunotoxicity studies on the cultivated clam *Tapes philippinarum* (Adams and Reeve, 1850). Mar Pollut Bull 39(1–12):112–115

Chiron S, Roy S, Cottier R, Jeannot R (2000) Speciation of butyl and phenyl-tin compounds in sediments using pressurized liquid extraction and liquid chromatography ICP-MS. J Chromatogr A 879:137–145

Clark EA, Sterrit RM, Lester JN (1988) Environmental science and technology. (22) 600 In: Dubey SK and Roy U. (2003) Biodegradation of tributyltins by marine bacteria. Appl Organ Chem 17:3–8

Cooke MG (2002) Effect of organotins on human aromatase activity in vitro. Toxicol Lett 126: 121–130

Davies B, Day J (1998) National Water Act news (various information pamphlets on the principles and the implementation of the new water act. http://www-dwarf.pwv.gov.za. Accessed online 1999 & 2000

De la Tone Fernando R, Ferrari L, Salibian A (2005) Biomarkers of a native fish species (*Cnesterodon decemmaculatus*) application to the water toxicity assessment of a peri-urban polluted water river of Argentina. Chemosphere 59(4):577–583

Delucchi F, Tombesi NB, Freije RH, Marcovecchio JE (2007) Butyltin compounds in sediments of the Bahia Blanca Estuary, Argentina. Environ Monit Assess 132:445–451

De Mora SJ, Stewart C, Philips D (1995) Sources and rate of degradation of tri (n-butyl) tin in marine sediments near Auckland, New Zealand. Mar Pollut Bull 30:50–57

Depledge MH, Fossi MC (1994) The role of biomarker in environmental assessment (2). Invert Ecotoxicol 3:161–172

Dowson PH, Bubb JM, Lester JN (1993) Temporal distribution of organotins in the aquatic environment: five years after the 1987 UK retail ban on TBT based antifouling paints. Mar Pollut Bull 26:482–494

Dubey SK, Roy U (2003) Biodegradation of tributyltins (organotins) by marine bacteria. Appl Organo Chem 17:3–8

EFSA (2004) Scientific panel on contaminants in the food chain, opinion on the health risks assessment to consumers associated with the exposure to organotins in foodstuffs. Question No EFSA- Q 2003-110. EFSA J 102:1–119. http://www.efsa.eu.int. Accessed 2004

EU- SCOOP (2006) Revised assessment of the risks to health and the environment associated with the use of organotin of the four organotin compounds (TBT, DBT DOT and TPT). Directorate general health and consumer protection. http://ec.europa.eu/health/ph-risk/commitees/04..../ scher_0_047.pdf. Accessed 30 Nov 2006

EVISA (European Virtual Institute for Speciation Analysis) (2010) Human exposure to organotin compounds via consumption of fish. http://www.speciation.net/public/info/contactform.html. 2010/03/08

Fanfandel M, Muller WEG, Batel R (2003) Molecular response to TBT stress in marine sponge (Suberites domuncula: proteolytical cleavage and phosphorylation of KRS_SD protein kinase. J Exp Mar Biol Ecol 297(2):239–252

Fent K (1996) Ecotoxicology of organotin compounds. Crit Rev Toxicol 26:1–117

Fent K, Hunn J (1995) Organotin in freshwater harbours and rivers: temporal distribution annual trends and fate. Environ Toxicol Chem 14:1123–1132

Fent K, Muller MD (1991) Occurrence of organotins in municipal wastewater and sewage sludge and behavior in a treatment plant. Environ Sci Technol 25:489–493

Forsyth DS, Jay B (1997) Organotin leachates in drinking water from chlorinated poly (vinyl chloride) (CPVC pipe). Appl Org Chem 11:551–558

FSA Food Standard Agency (2005) Survey of organotins in shellfish. http://www.food.gov.uk/multimedia/pdfs/fsis8105.pdf. Accessed

Gibbs PE, Bryan GW (1996) Reproductive failure in the gastropod nucella lapillus associated with imposex caused by tributyltin pollution: a review. In: Champ MA, Seligman PF (eds) Organotin environmental fate and effects. Chapman and Hall, London, pp 259–281

Gil R, Avital G, Stewart I, Yehuda B (2000) Unregulated use of TBT-based antifouling paints in Israel: high contamination and imposex levels in two species of marine gastropods. Mar Ecol Prog Ser 192:229–239

Haggera JA, Depledge MH, Galloway TS (2005) Toxicity of tributyltin in the marine mollusk *Mytilus edulis*. Mar Pollut Bull 51:811–816

Harino H, Fukushima MH, Kawai S (2000) Accumulation of butyltin and phenyltin compounds in various fish species. Arch Environ Contam Toxicol 39:13–19

Harino H, Iwasaku N, Arai T, Ohji M, Miyazaki N (2005) Accumulation of organotin compounds in the deep- sea environment of Nankai Trough, Japan. Arch Environ Contam Toxicol 49: 497–503

Harino H, Madoka O, Wattayakorn G, Adulyyanukosol K, Arai T, Miyazaki N (2008) Accumulation of organotin compounds in tissues and organs of dolphins from the coasts of Thailand. Archi Environ Contam Toxicol 54:146–153

Harino H, Ohji M, Wattayakorn G, Arai T, Rungsupa S, Muyazaki N (2006) Occurrence of antifouling biocides in water and sediments from the port of Osaka, Japan. Arch Environ Contam Toxicol 48(3):303–310

Hawkins SJ, Gibbs PE, Pope ND, Burt GR, Chessman BS, Bray S, Proud SV, Spence SK, Southward AJ, Southward GR, Langston WJ (2000) Recovery of polluted ecosystems: the case for long-term studies. Mar Environ Res 54(3–5):215–222

Hermosin MC, Martin P, Cornejo J (1993) Adsorption mechanisms of monobutyltins in clay minerals. Environ Sci Technol 27:2606–2611

Hoch M (2001) Organotin compounds in the environment- an overview. Appl Geochem 16: 719–743

Hongwen S, Guolan H, Shugui D (1996) Adsorption behaviour and OSPR studies of organotin compounds on estuarine sediments. Chemosphere 33(5):831–838

Horiguchi T, Imai T, Cho HS, Shiraishi H, Slubata Y, Morita M (1998) Acute toxicity of organotin compounds to the larvae of the rock shell Thais clavigera, the disk abalone, haliotis madaka. Mar Environ Res 46:469–473

Horiguchi T, Li Z, Uno S, Shimizu M, Shirashi V, Morita M, Thompson JAJ, Lerings CD (2004) Contamination of organotin compounds and imposex in mollusc from Vancouver, Canada. Mar Environ Res 52(1–2):75–88

Hu J, Zhen H, Wan Y, Gao J, An W, An L, Jin P, Jin X (2006) Tropic magnification of triphenyltin in a marine food web of Bohai Bay, North China: comparison to TBT. Environ Sci Technol 40(10):3142–3147

Huggett RJ, Unger MA, Seligman PF, Valkirs AO (1992) The marine biocide tributyltin: assessing and managing the environmental risks. Environ Sci Technol 26:232–237

International Maritime Organisation (IMO) (2001) International convention on the control of the harmful antifouling systems on ships. http://iyp.yachtpaint.com/usa_old/superyacht/default.asp. Accessed 11 Jan 2007

Isani G, Andreau G, Kindt M, Carperie E (2000) Metallothioneins (MTs) in marine mollusks. Cell Mol Biol 46(2):304–313

Iwata H, Tanabe S, Mizuno T, Tatsukawa R (1995) High accumulation of toxic butyl-ins in marine mammals from Japanese coastal waters. Environ Sci Technol 29:2959–2962

Kannan K, Lee RF (1996) Triphenyltin and its degradation products in foliage and soils from sprayed pecan orchards and in fish from adjacent ponds. Environ Toxicol Chem 15:1492–1499

Konstantinou IK, Albanis TA (2004) worldwide occurrence and effects of antifouling paint booster biocides in the aquatic environment: a review. Environ Int 30:235–248

Kuballa J, Jantzen E, Wilken RD (1996) Organotin compounds in sediments of the rivers Elbe and Muide. In: Calmano W, Forstner U (eds) Sediments and toxic substance, Environmental effect and ecotoxicity. Springer, Berlin, pp 245–270

Langston WJ, Popoe ND (1995) Determination of TBT absorption and desorption in estuarine sediments. Mar Pollut Bull 31(1–3):32–43

Largardic L, Caquet T, Ramade F (1994) The role of biomarkers in environmental assessment. Invertebrates Popu Comm Ecotoxi 3:193–208

Laughlin RB (1996) Bioaccumulation of TBT by aquatic organisms. In: Champ MA, Seligman PF (eds) Organotin environmental fate and effects. Chapman and Hall, London, pp 331–357

Leinio S, Lehtonen KK (2005) Seasonal variability in biomarkers in the bivalves *Mytilus edulis* and *Macoma balthica* from the northern Baltic Sea. Comp Biochem Physiol C Toxicol Pharmacol 140(3–4):408–421

Lionetto MG, Giord nano ME, Caricato R, Pascariello MF, Marinosci L, Shettino T (2001) Biomonitoring of heavy metals contamination along the Salento Coast (Italy) by metallothionein evaluation in *Mytilus galloprovincialis* and *Mullus barbatus*. Aquatic Conserv Mar Freshw Ecosyst 11:305–310

Lisicio C, Carro DM, Magi E (2009) Comparison of two analytical methods for the determination of organotin compounds in marine organisms. Comptes Rendus Chimie 12:831–840

Lo S, Allera A, Alberts P, Heimbrechi J, Janzen E, Klingmuller D, Stockclbroeck S (2003) Dithioerythritol (DTE) prevents inhibitory effects of TPT on the key enzymes of the human sex steroid hormone metabolism. J Steroid Biochem 41:1560–1565

Magni P, De Falco G, Falugi C, Frauzoui M, Monteverdi M, Perrone E, Squo M, Bolognesi C (2006) Genotoxicity biomarkers and acetylcholinesterase activity in natural populations of *Mytilus galloprovincialis* along a pollution gradient in the Gulf of Oristano. Environ Pollut 142(1):65–72

Maguire RJ (1996) The occurrence, fate and toxicity of tributyltin and its degradation products in fresh water environments. In: de Mora SJ (ed) Tributyltin: case study of an environmental contaminant. Cambridge University Press, Cambridge, pp 94–98

Marshall DJ, Rajkumar A (2003) Imposex in the indigenous *Nassarius kraussianus* (Mollusca: *Neogastropoda*) from South African harbours. Mar Pollut Bull 46:1150–1155

Matozzo V, Ballarin L, Marin MG (2002) In vitro effects of tributyltin on functional responses of haemocytes in the clam *philippinarum*. Appl Org Chem 16:169–174

Matozzo V, Tomei A, Marin MG (2005) Acetylcholinesterase as a biomarker of exposure to neurotoxic compounds in the clam *Tapes philippinarum* from the Lagoon of Venice. Mar Pollut Bull 50(12):1686–1693

Maureen EC, Willingham GL (1996) Biodegradation of TBT by marine bacteria. Biofueling 10:239

Meador JP, Rice CA (2001) Impaired growth in the polychaete *Armandia brevis* exposed to TBT in sediments. Mar Environ Res 51:43–29

Meng-Pei H, Shin-Mei L (2003) Accumulation of organotin compounds in Pacific oysters, *Crassostrea gigas*, collected from aquaculture sites in Taiwan. Sci Total Environ 313:41–48

Micic M, Bihari N, Labura Z, Muller WEG, Batel R (2001) Induction of apoptosis in the mussel *Mytilus galloprovincialis* by tri-n-butyltin chloride. Aquat Toxicol 55:61–73

Morabito R, Quevauviller P (2002) Performances of spectroscopic methods for TBT determination the 10 years of the EU-SM and T organotin programme. Spectrosc Europe 14:18–28

Mourgand Y, Martinez E, Geffard A, Andradal B, Stanisiere JY, Amiard JC (2002) Metallothionein concentration in the mussel *Mytilus galloprovincialis* as a biomarker of response to metal contamination: validation in the field. Biomarkers 7(6):479–490

Nakanish T (2007) Potential toxicity of organotin compound via nuclear receptor signaling in mammals. J Health Sci 53:1–9

Nicholson S, Lam PKS (2005) Pollution monitoring in Southeast Asia using biomarkers in the mytilid mussel *Perna viridis* (Mytilidae: Bivalvia). Environ Int 31:121–132

Okoro HK, Fatoki OS, Adekola FA, Ximba BJ, Snyman RG (2011) Sources, Environmental levels and toxicity of organotin in marine environment – a review. Asian J Chem 23(2):473–482

Olsson PE, Kling P, Hongstrand C (1998) Mechanisms of heavy metal contamination and toxicity in fish. In: Langston WJ, Bebianno MJ (eds) Metal metabolism in aquatic environments. Chapman and Hall, London, pp 321–350

Pann R, Anu T, Pia K, Verkasalo K, Satu M, Terttu V (2008) Blood levels of organotin compounds and their relation to fish consumption and Finland. Sci Total Environ. http://dx.doi.org/10.101b/ J Scitotenv 399(1–3):90–95

Paton GI, Cheewasedtham IL, Dawson M (2006) Degradation and toxicity of phenyltin compounds in soil. Environ Pollut 144:746–751

Peachery RBJ (2003) Tributyltin and polycyclic aromatic hydrocarbon levels in Mobile Bay Alabama: a review. Mar Pollut Bull 41:1305–1371

Pessoa MF, Fernando A, Oliveira JS (2001) Use of imposex (pseudohermaphroditism) as indicator of the occurrence of organotin compounds in Portuguese coastal waters – sado and mira estuaries, Vol. 16(3). John Willey & Sons, Inc., pp 234–241

Petrovic S, Ozretic B, Krajnovic-Ozretic M, Bobinac D (2001) Lysosomal membrane stability and metallothionein in digestive gland of mussels (*Mytilus galloprovincialis* Law) as biomarkers in a field study. Mar Pollut Bull 42(12):1373–1378

Pfeiffer s, Doris S, Dippner JW (2005) Effects of temperature and salinity on acetylcholinesterase activity, a common pollution biomarker, in *Mytilus sp*. From the south- western, Baltic Sea. J Exp Mar Biol Ecol 320(1):93–103

Raspor B, Dragun Z, Erk M, Ivankovie D, Pavii J (2004) Is the digestive gland of *Mytilus galloprovincialis* a tissue of choice for estimating cadmium exposure by means of metallothioneins? Sci Total Environ 333(1–3):99–108

Rebeiro CAO, Schwartzman M, Silva de Assis HC, Silva PH, Pelletier E, Alkaishi M (2002) Evaluation of TBT subchronic effects in tropical freshwater fish (*Astyanax bimaculatus*, Linnaeus, 1758). Ecotoxicol Environ Saf 51(3):161–167

Rees CM, Brachy BA, Fabus GJ (2001) Incidence of imposex in thais orbital from port Phillip Bay (Victoria, Australia), following 10 years of regulation on use of TBT. Mar Pollut Bull 42(10):873–878

Rodriguez-Ortega MJ, Alhama T, Fues V, Romero-Ruiz A, Rodriguez-Ariza A, Lopez-Barea J (2002) Biochemical's biomarkers of pollution in the clam chamelea gallina from South-Spanish littoral. Environ Toxicol Chem 21:542–549

Roeva NN, Sidrov AV, Ymovitskii YG (1999) Metallothioneins, proteins binding heavy metals in fish. Bio Bull 26(6):617–622

Ross KS, Cooper N, Bidwell JR, Elder J (2002) Genetic diversity and metal tolerance of two marine species: comparisons between populations from contaminated and reference sites. Mar Pollut Bull 44:671–679

Ruiz JM, Bryan GW, Gibbs PE (1995) Effects of TBT exposure on the veliger larvae development of the bivalve *Scrobicularia plana* (da Costa). J Exp Mar Biol Ecol 9186:53–63

Sadiki AI, Williams DT (1999) A study on organotin levels in Canadian drinking water distributed through PVC Pipe. Chemosphere 38:1541–1548

Sarkar A, Ray D, Amulya NS, Subhodeep S (2006) Molecular biomarkers; their significant and application in marine pollution monitoring. Ecotoxicology 15:333–340

Sarradin PM, Lapaquellerie Y, Astruc A, Lactouche C, Astruc M (1995) Sci Total Environ 170:59

Schroth W, Ender A, Schierwater B (2005) Molecular biomarkers and adaptation to environmental stress in moon jelly (*Aurelia spp.*). Mar Biotechnol 7:449–461

Sentosa AP, Bachri M, Sadijah A (2009) Speciation of organotin compounds with ion- pair-reversed phase chromatography technique. Eurasian J Anal Chem (EJAC) 4(2):215–225

Sergi D, Silva L, Panta V, Barcelo D, Josep MB (2005) Survey of organotin compounds in rivers and coastal environment in Portugal 1999–2000. Environ Pollut 136:525–586

Smith BS (1981) Tributyltin compounds induce male characteristics on female mud Snails *Nassarius obsoletus* (*Say*). J Appl Toxicol 1:141–144

Stab JA, Tras TP, Strounberg G, Van Kesteren JS, Leonands P, Van Hattum B, Brickman UAT, Cofino WP (1996) Determination of organotin compounds in the food webs of a shallow freshwater lake in the Netherlands. Arch Environ Contam Toxicol 31:328

Strand J, Larsen MM, Lockyer C (2005) Accumulation of organotin compounds and mercury in harbor porpoises (*Phocoena phocoena*) from the danish waters and West Greenland. Sci Total Environ 350:59–71

Sun H, Guolan H, Shugui D (1996) Adsorption behavior and OSPR studies of organotin compounds on estuarine sediments. Chemosphere 33(5):831–838

Takahashi S, Mukai H, Tanabe S, Sakayana K, Miyazaki T, Masuno H (1999) Butyltins residues in livers of humans and wild terrestrial mammals and in plastic products. Environ Pollut 106: 213–218

Takahashi S, Tanabe S, Kubodera T (1997) Butyltin residues in deep-sea organisms collected from Suruga Bay, Japan. Environ Sci Technol 31:3103–3109

Takashi S, Kazunari K, Mitsumu U, Mitsuru M (1997) Chemical species of organotin compounds in sediments at a Marina. J Agric Food Chem 47:3886–3894

Tanabe S, Prudente M, Mizuno T, Hasegawa J, Iwata H, Miyazaki N (1998) Butyltin contamination in marine mammals from the North Pacific and Asian coastal waters. Environ Sci Technol 32:193–198

Tsuda T, Inone T, Kojima M, Aoki S (1995) Daily intakes of tributyltins and triphenyltin compounds from metals. J ADAC Int 78:941–943

US Environmental Protection Agency (2001) Office of the Air quality planning and standards Bioaccumulation summary – Tributyltin. http://www.epa.gov/ttu/uatw/hithef/eth-phth.html

US United State Code (1988) Organotin paint controls. http://uscodehouse.gov/download/33C37. tx. Accessed 2007

Valkirs AO, Seligman PF, Haslbeck E, Caso JS (2003) Measurements of copper release rates from antifouling paint under laboratory and in situ conditions: implications for loading estimation to marine water bodies. Mar Pollut Bull 46:763–779

Van-Gestel CAM, Van Brummelen TC (1996) Incorporation of the biomarkers concept in ecotoxicology calls for a redefinition of terms. Ecotoxicology 5:217–225

Visoottiviseth P, Kruawan K, Bhumiratana A, Wileirat P (1995) Isolation of bacterial culture capable of degrading triphenyltin pesticides. Appl Org Chem 9:1–9

Wadlock MJ, Thain JE (1983) Shell thickening in *Crassostrea gigas*. Organotin antifouling or sediment induced. Mar Pollut Bull 14:411–415

Wagner G (1993) Plants and soils as specimen types from terrestrial ecosystems in the environmental specimen banking program of the Federal Republic of Germany. Sci Total Environ 139–140, 213–224

WHO-IPCS (1999) World health organization. international programme on chemical safety. Tributyl compounds. Environ Health Criteria 116. http://www.inchem.org/documentsehc/ehc/ehc/116.htm. Assessed 2007

Worldwide fund for Nature (1995) Marine pollution by triorganotins. Marine update WWF UK Godalining, p 4

Zhang G, Yan J, Fu JM, Parker A, Li XD, Wang ZS (2003) Butyltins in sediments and biota from the Pearl River Delta, South China. Chem Spec Bioavail 14:35–42

Shellfish and Residual Chemical Contaminants: Hazards, Monitoring, and Health Risk Assessment Along French Coasts

Marielle Guéguen, Jean-Claude Amiard, Nathalie Arnich,
Pierre-Marie Badot, Didier Claisse, Thierry Guérin,
and Jean-Paul Vernoux

Contents

J.-P. Vernoux (✉)
Unité des microorganismes d'intérêt laitier et alimentaire EA 3213, UFR ICORE 146,
Université de Caen-Basse Normandie, 14032 Caen Cedex 5, France
e-mail: jean-paul.vernoux@unicaen.fr

D.M. Whitacre (ed.), *Reviews of Environmental Contamination and Toxicology*,
Reviews of Environmental Contamination and Toxicology 213,
DOI 10.1007/978-1-4419-9860-6_3, © Springer Science+Business Media, LLC 2011

1 Introduction

Shellfish farming is a common industry along European coasts. According to the 2005–2006 data from the French National Shellfish Farming Committee (CNC – *Comité National de la Conchyliculture* 2010; see Table 1 for a list of acronyms and abbreviations used in this chapter). Spain is the largest shellfish producer in Europe (~270,000 t) and France ranks second, producing 200,000 t of shellfish annually. France is the leading European oyster producer, with an annual output of 130,000 t of *Crassostrea gigas,* and ranks fourth in the world after China, Japan, and Korea. The top three European mussel (*Mytilus edulis* and *Mytilus galloprovincialis*) producers are Spain (260,000 t), Denmark (80,000 t), and France (65,000 t). For other shellfish, the French annual output level is 15,000 t for king scallops (*Pecten maximus*) and a few thousand tons for *Ruditapes* clams (*Ruditapes decussatus* and *Ruditapes philippinarum*) and cockles (*Cerastoderma edule*). The economic impact of shellfish farming is considerable; despite fairly long production lead times and difficult operating conditions, shellfish farming generates annual sales of more than 650 million Euros in France, owing to its high added value.

The main species of shellfish consumed in France are the Pacific oyster (*C. gigas*), mussels (*M. edulis* and *M. galloprovincialis*), king scallop (*P. maximus*), winkle (*Littorina littorea*), whelk (*Buccinum undatum*), cockle, *Ruditapes* clams, and scallops (*Pecten* spp., *Chlamys* spp.). All of these species play a prominent role in French diets and in festive customs. But these species sometimes produce acute food poisoning in consumers from phycotoxins (AFSSA 2008c) that the shellfish ingest through planktonic microalgae, particularly dinoflagellates, or from ingesting microbes (bacteria and viruses). Mineral and organic chemical contaminants of human origin (referred to below as residual chemical contaminants) can also accumulate in shellfish and potentially cause chronic poisoning (Bügel et al. 2001; Mozaffarian and Rimm 2006). Accordingly, bivalve mollusks are known to be reliable indicators of the marine environment, because they accumulate many anthropogenic pollutants (Goldberg 1975; Goldberg et al. 1978; Vos et al. 1986).

Current European regulations focus on regulating microbiological agents, phycotoxins, and some chemical contaminants. Since 2006, these regulations have been compiled under the name of the "'Hygiene Package." Because of increasing concern for the presence of contaminants in the marine environment, the French Food Safety Agency (AFSSA; now named the French Agency for Food, Environmental and Occupational Health & Safety, ANSES) issued a report in 2008 on the monitoring of chemicals in shellfish-farming areas and on health risks associated with shellfish consumption (AFSSA 2008b).

The purpose of this review is to address the residual chemical hazards that exist in shellfish that are routinely sampled from the natural marine environment and from the market place. We have included data on exposure levels and body burdens of many contaminants, and have related these data to human health risks. We have also addressed the concentration of contaminants found in the context of current

Table 1 List of abbreviations and acronyms used in this review

AFSSA: *Agence Française de Sécurité Sanitaire des Aliments* (French Food Safety Agency) (Web site: www.anses.fr)

ANSES: *Agence Nationale de Sécurité Sanitaire de l'alimentation, de l'environnement et du travail* (French Agency for Food, Environmental, and Occupational Health & Safety) (Web site: www.anses.fr)

ATSDR: Agency for Toxic Substances and Disease Registry

BCF: bioconcentration factors

BWT: body weight

$BMDL_{01}$: benchmark dose (lower confidence limit 0.01)

$BMDL_{05}$: benchmark dose (lower confidence limit 0.05)

BQSPMED: *Bureau de la Qualité Sanitaire des Produits de la Mer et d'Eau Douce* (Office for the Quality and Safety of Food Products from Fresh and Marine Waters)

BRAB: Bureau de la Réglementation Alimentaire et des Biotechnologies (Office of Food and Biotechnology Regulations)

BTEX: benzene, toluene, ethylbenzene, xylene

CALIPSO: Etude des Consommations ALimentaires de produits de la mer et Imprégnation aux éléments traces, PolluantS and Omega-3 (fish and seafood consumption study and biomarker of exposure to trace elements, pollutants, and omega-3)

CF: concentration factor

CNC: French national shellfish-farming committee

DBT: dibutyltin

DDAM: Direction Départementale des Affaires Maritimes (local Maritime Affairs Authorities)

DDT: dichlorodiphenyltrichloroethane

DDE: dichlorodiphenyldichloroethylene

DDD: dichlorodiphenyldichloroethane

DDSV: Direction Départementale des Services Vétérinaires (local veterinary authorities)

DEHP: di(2-ethylhexyl)phthalate

DGAL: Direction Générale pour l'Alimentation (French Directorate for Food)

DGS: *Direction Générale de la Santé* (French Directorate General for Health)

DMA: dimethylarsinic acid

DOT: dioctyltin

DPMA: *Direction des Pêches Maritimes et de l'Aquaculture* (Directorate for Marine Fisheries and Aquaculture)

DPT: diphenyltin

EAT: Etudes Alimentaires Totales (total diet study (TDS))

EC: European Community

EEC: European Economic Community

EFSA: European Food Safety Authority

EPA: Environmental Protection Agency

EU-RL: EU reference laboratory

FAO: Food and Agriculture Organization of the United Nations

GST: glutathione S-transferase

HACCP: Hazard Analysis Critical Control Point

IAEA: International Atomic Energy Agency

IARC: International Agency for Research on Cancer

IFREMER: *Institut Français de Recherche pour l'Exploitation de la Mer* (French Research Institute for Exploitation of the Sea)

INCA: Enquête Individuelle et Nationale sur la Consommation Alimentaire (consumption data for the general population)

INRS: *Institut National de Recherche et de Sécurité* (National Institute of Research and Safety)

IRSN: *Institut de radioprotection et de sûreté nucléaire* (French Institute for Radiation Protection and Nuclear Safety)

Table 1 (continued)

JECFA: Joint FAO/WHO Expert Committee on Food Additives
JORF: *Journal Officiel de la république Française* (Official Journal of the French Republic)
LD_{50}: lethal dose 50%
LERQAP: Laboratoire d'Etudes et de Recherches sur la Qualité des Aliments et les Procédés
    Agroalimentaires (Laboratory of studies and research on food quality and food processes)
MAP: Mediterranean Action Plan
MCSI: *Mission de Coordination Sanitaire Internationale* (International Health and Safety
    Coordination Mission)
MeHg: methylmercury
MED POL: Barcelona Convention for the Protection of the Mediterranean Sea Against Pollution
MBT: monobutyltin
MMA: monomethylarsonic acid
MOREST: *Mortalité ESTivale d'Huîtres* (oyster summer mortality program)
MPT: monophenyltin
MT: metallothioneins
NPE: nonylphenol ethoxylates
NRL: National Reference Laboratory
OCA-EN: Observatoire des Consommations Alimentaires-Epidémiologie Nutritionnelle (Food
    Consumption and Nutritional Epidemiology Unit)
OPE: octylphenol ethoxylate
OSPAR: Convention for the protection of the marine environment of the North-East Atlantic
P95: 95th percentile
PAH: polycyclic aromatic hydrocarbon
PCB: polychlorinated biphenyls
PCBi: indicator PCBs (sum of selected PCBs)
DL-PCB: PCB dioxin-like
PCDD/Fs: polychlorinated dibenzo-dioxins/furans
PTMI: provisional tolerable monthly intake
PTWI: provisional tolerable weekly intake
REPAMO: *Réseau de Pathologie des Mollusques* – Mollusk pathology network
RNO: *Réseau National d'Observation* – French National Monitoring Network
ROCCH: Réseau d'Observation de la Contamination CHimique du milieu marin (French
    National Monitoring Network)
SCOOP: Scientific Cooperation
TBT: tributyltin
TDI: tolerable daily intake
THg: total mercury
TPT: triphenyltin
TWI: tolerable weekly intake
UNEP: United Nations Environment Programme
WFD: Water Framework Directive
WHO: World Health Organization
WT: Weight

regulatory and food safety standards. The data compiled here are designed to provide readers with a basis for assessing whether or not it is necessary to continue or even extend environmental chemical contaminant monitoring to other chemicals that pose significant potential consumer health risks.

2 Regulation of Shellfish Food Safety in Europe

Food safety monitoring of shellfish-farming areas falls under European regulatory jurisdiction and is defined in the "Hygiene Package", which came into force on 1 January 2006. There are several specific sections of this regulation that apply to live bivalve mollusks. Two of these regulations (EC 2004a, b) are directed toward industry professionals (No. 852/2004 and No. 853/2004), and two others (EC 2004c; 2006b) apply to competent authorities having to do with official controls (No. 854/2004 and No. 882/2006). Directive (EEC) No. 492/91 (EEC 1991), which had previously set the hygiene rules for the production and marketing of live bivalve mollusks, was repealed. A general presentation of these regulations is presented below and deals only with the sections on residual chemical contaminants.

2.1 Provisions of the Hygiene Package

Regulation (EC) 852/2004 (EC 2004b) lays down general rules on food hygiene, and applies to primary production (farm and fishery products). It is complemented by Regulation (EC) 853/2004 (EC 2004a), which lays down additional specific hygiene rules for products of animal origin. Annex III, Section VII of Regulation (EC) 853/2004 specifies the requirements for live bivalve mollusks. Regulations (EC) 854/2004 and 882/2006 (EC 2004c; 2006b) apply to official control bodies and define a legal framework for setting the locations and boundaries of production, and relaying areas (depurating areas). The regulations also require food safety monitoring, by sampling, to screen for chemical and microbiological contaminants.

A clear distinction must be made between primary production of shellfish and the other operations that are required to bring shellfish to the market, because the regulatory obligations are different. Primary shellfish production concerns all operations carried out before shellfish reach an approved purification establishment: rearing, harvesting, and transport of the produce. Annex I of Regulation (EC) 852/2004 and some provisions in Annex III, Section VII of Regulation (EC) 853/2004 apply to primary producers. Producers must be registered but are under no obligation to set up Hazard Analysis Critical Control Point (HACCP) procedures. The activities of the purification and dispatch establishments (finishing, packing, etc.) are not regarded to constitute primary production. The provisions of Annex II of Regulation (EC) 852/2004 and of Annex III, Section VII of Regulation (EC) 853/2004 apply to these establishments. These establishments must be approved by the competent authority and are under an obligation to introduce HACCP procedures.

The classification of production into Class A, B, and C areas is based solely on measures having to do with microbiological contamination; these measures are defined by the Hygiene Package, and Regulations 853/2004 and 854/2004 in particular:

Class A areas are those from which live bivalve mollusks may be harvested for direct human consumption;

Class B areas are those from which live bivalve mollusks approach conformity, but before being marketed for human consumption, they require a short but sufficient purifying treatment;

Class C areas are those from which live bivalve mollusks can be harvested only after relaying (depurating) for a long period, with purification, or after intensive purification by an appropriate method.

At the EC level, the Hygiene Package regulates the monitoring of production areas during operations (854/2004, Annex II, Chapter II.b) for three types of hazards: microbiological, phytoplanktonic/phycotoxic, and chemical. Thus, although under the Hygiene Package, there is no obligation to test for chemical contaminants for the purposes of classifying the production areas and there is an obligation to chemically monitor these areas.

2.2 Provisions on Chemical Contaminants

To be regarded as edible, bivalve mollusks must also comply with maximum levels of certain contaminants defined in Regulation (EC) 1881/2006 of 19 December 2006 (EC 2006c), which replaces Regulation (EC) 466/2001 (EC 2001), as amended by Regulation (EC) 629/2008 of 2 July 2008 (EC 2008a). These contaminant thresholds (Table 2) apply to the edible parts of bivalve mollusks, i.e., the whole flesh, except for the king scallop, for which the digestive gland is not taken into account (Article 1 of Regulation (EC) 1881/2006). Non-bivalve mollusks (gastropods), echinoderms, and tunicates are not covered by the European regulations, but in France, in a recommendation issued on 31 October 2007 (AFSSA 2007b), AFSSA considers that the cadmium threshold set by decree on 21 May 1999 is appropriate: 2 mg kg^{-1} fresh mass for whelks (gastropod, *B. undatum*) (JORF 1999). For echinoderms and tunicates, given their particularly low levels of consumption, it is regarded as not necessary to set a regulatory threshold, but rather a guideline value of 2 mg kg^{-1} fresh mass was set (AFSSA 2007b).

Table 2 Regulatory thresholds for consumption of various contaminants in bivalve mollusks (EC 2006c, amended by EC 2008a)

Contaminant		Maximum level (fresh wt)
Metals	Lead	1.5 mg kg^{-1}
	Cadmium	1 mg kg^{-1}
	Mercury	0.5 mg kg^{-1}
Dioxins and PCBs	Dioxins	4 pg g^{-1}
	Dioxins + DL-PCBs	8 pg g^{-1}
PAHs	Benzo[*a*]pyrene	10 μg kg^{-1}

3 Identifying Residual Chemical Hazards in the Marine Environment and in Shellfish

To identify the risks of chemical residues in the marine environment being transferred to bivalve mollusks, and thence to humans, it is necessary to target, among the many potentially toxic chemicals, those that have a likelihood of being released by human activities in the vicinity of shellfish-farming areas. That does not mean that contamination of the environment and of the bivalve mollusks by the chemicals addressed in this chapter has always been demonstrated. Hazard identification is usually conducted independently of the likelihood of an accident occurring. Consequently, hazard identification does not include addressing chemicals that may be released into the environment from hitherto unidentified sources or following accidental spills, irresponsibility, or acts of malice.

The main sources of contaminants are of human origin (Manta et al. 2002). They involve the following: terrestrial and marine crop and livestock farming; human habitation (energy production, building and demolition, wastewater, incineration of household waste, heating, etc.); land transport (infrastructures and vehicles); energy production; industry (solid waste, liquid effluents and gas emissions, end-of-life products, etc.); maritime transport and related activities (port activities, dredging, etc.), as well as some leisure activities (golf courses, water sports, sailing, etc.). Moreover, pharmaceutical residues have been found in environmental waters and in the marine environment, so they also could qualify as pollutants of interest (Walraven and Laane 2009; Fatta-Kassinos et al. 2011).

Crop and livestock-farming activities result in the release of organic matter and nutrients (nitrates, phosphates, and potassium) into the environment; these can contribute to the eutrophication of the marine environment and cause major changes to aquatic community dynamics. Many chemicals are or have been used in farming: plant protection products, biocides, veterinary drugs (including antibiotics), any of which may contaminate the marine environment at some time (Schaffner et al. 2009). Human habitations can also be major sources of organic matter release into aquatic environments, particularly in coastal areas, via wastewater release (Heinzow et al. 2007; Schaffner et al. 2009). Incinerators and domestic heating equipment release persistent organic pollutants (POPs), such as dioxins, PCBs, and PAHs (Lewtas 2007; Van Caneghem et al. 2010). Industrial activities also release a very wide range of toxic chemicals. Transport and energy production release such substances as PAHs, trace elements, radionuclides, and many atmospheric pollutants (England et al. 2001). Through their toxic potential, these substances can cause direct adverse effects on the marine environment and on farmed mollusks, and indirect effects on human consumers.

3.1 Inorganic Contaminants

Metals (trace elements) are naturally present in many rocks and minerals. Due to natural weathering of the earth's crust, they are found in all environmental compartments, including seawater. Some trace elements that are absorbed by living organisms accumulate in the food chain and therefore present a risk to humans, who are the final consumers at the top of the food chain (Hamilton 2004; Hillwalker et al. 2006). Shellfish filter large amounts of water to extract their food and are excellent bioaccumulators (Claisse 1989). Any contaminants in the water, from natural sources or pollution, are easily concentrated in shellfish flesh, particularly metals, such as the following: mercury, cadmium, lead, copper, and zinc. Metals are mainly fixed in particular organs, such as the digestive gland (Soto et al. 1996), which plays a part in assimilation, excretion, and detoxification (Johnson et al. 1996). These organs are generally the parts of the organisms that are eaten by humans (except for king scallops whose flesh is consumed only in France).

In Tables 3 and 4, we summarize the main metal contaminants found in the environment, their human–activity sources, and we categorize their toxicity and risk levels. Levels of contamination in marketed shellfish are given by species for the three regulated metal contaminants (lead, cadmium, and mercury); the results come from the CALIPSO (2005) and first total diet study (EAT 2004) which were performed in France (Table 3). The levels reported in these tables can be compared with the maximum permitted levels set for fishery products. For example, cadmium levels are above the maximum permitted limits in some scallop species (1.14 mg kg^{-1} fresh wt), while the other bivalve mollusks show lower levels – no more than 0.040 mg kg^{-1} fresh wt. For lead and mercury, none of the species sampled were above the maximum permitted levels (lead < 0.26 mg kg^{-1} fresh wt and mercury < 0.003 mg kg^{-1} fresh wt). The observed values in French shellfish-farming areas (Fig. 1a, b, c and e) are very close to those observed in marketed shellfish just before consumption.

Table 3 also shows that mollusks have high concentrations of arsenic, the highest levels being found in whelks (15.8 mg kg^{-1} fresh wt). However, contamination levels in shellfish are lower than those in crustaceans, fish, and other seafood; the highest levels were found in octopus (42 mg kg^{-1} fresh wt; Leblanc et al. 2006; Sirot et al. 2009). In 1988, the mean arsenic levels in bivalve mollusks (mussels and oysters) along the French coast ranged from 10 to 30 mg kg^{-1} (Michel 1993); arsenic residues were the most frequently encountered, irrespective of geographical area and species. It is difficult to link the highest levels with possible pollution sources. For example, organisms in the major estuaries (Seine, Loire, and Gironde rivers) are less contaminated than those in adjacent coastal areas. It seems that the levels of arsenic in the environment derive less from bioaccumulation than from whether the metal is in organic or inorganic form (Michel 1993). In laboratory experiments, the oyster *Crassostrea virginica* bioaccumulates little inorganic arsenic and only a fraction of the organic arsenic present in the phytoplankton (Sanders et al. 1989). The arsenic fixed on inert particles of seston is poorly bioconcentrated in the oyster *C. gigas* (Ettajani et al. 1996), but the small amount that passes through the oyster causes intense erosion of the mitochondrial cristae, leading eventually to cellular respiratory failure. In the peppery furrow shell (or sand gaper) *Scrobicularia plana*, bioconcentrated arsenic levels match the levels of sediment contamination (Langston 1983). In the winkle, arsenic levels vary from 9 to 70 mg kg^{-1} dry wt, their exact level depending on the degree of contamination of their food sources

Table 3 Levels of contamination in environment and in shellfish flesh sampled from the marketplace for three inorganic contaminants (Cd, cadmium; Pb, lead; Hg, mercury) regulated according to EC (2006c) and amended by EC (2008a). Arsenic (As), though not regulated, is included in the table, because it is also closely monitored

	Cd	Pb	Hg	As
Anthropogenic source[c]	Industry (coloring, stabilizer, and cadmium plating)	Industry (printing. metallurgy, etc.)	Rare in the natural environment, electrical industry, etc.	Rare in the natural environment, metallurgy industries, etc.[c]
Mean levels in the environment				
Seawater (μg L^{-1})[a]	0.01–0.1	0.5–5	0.005–0.05	1–2[a]
Sediments (μg g^{-1} dry wt)[a]	0.1–1	5–50	0.05–0.5	5–3000[a]
Contamination in shellfish (mg kg^{-1} fresh wt)				
Regulatory threshold	1[d]	1.5[d]	0.5[d]	
Oyster (min–max) ($n = 6$)*	0.07–0.22[d]	0.04–0.08[d]	0.003–0.02[d]	0.003[d]
Mussel (min–max) ($n = 6$)*	0.06–0.18[d]	0.14–0.26[d]	0.003–0.02[d]	0.88–3.39[d]
Cockle (mean) ($n = 2$)**	0.04[b]	0.04[b]	0.02[b]	1.78[b]
Scallop ($n = 1$)**	1.14[b]	0.09[b]	0.01[b]	2.42[b]
Winkle (mean) ($n = 3$)**	0.19[b]	0.09[b]	0.01[b]	6.39[b]
Whelk (mean) ($n = 3$)**	0.78[b]	0.06[b]	0.03[b]	15.8[b]
King scallop (mean) ($n = 4$)**	0.27[b]	0.07[b]	0.03[b]	2.96[b]
PTWI (μg kg^{-1} bwt week^{-1})	7[b] (P1) 2.5[e] (P2)	25[b] (P4) withdrew in 2010[g]	1.6 (MeHg) and 5 (Hg total)[b] (P5) 4 (Hg inorganic)[f] (P6)	Intake not to exceed[b]: 15 (As^{III} and As^{V}) and 350 (total As) Withdrawn in 2010[f]
PTMI (μg kg^{-1} bwt month^{-1}) with (Px) = PTWI or PTMI values	25[g] (P3)			

Table 3 (continued)

	Cd	Pb	Hg	As
Anthropogenic source[c]	Industry (coloring, stabilizer, and cadmium plating)	Industry (printing. metallurgy, etc.)	Rare in the natural environment, electrical industry, etc.	Rare in the natural environment, metallurgy industries, etc.[c]
Contribution of shellfish in % of PTWI or PTMI (Px)				
High consumers (CALIPSO)[b]	8.22% (P1) – 23% (P2) – 10%(P3)	0.8% (P4)	0.12%(MeHg) (P5) – 1.16% (P6)	1.2% (As^{III} and As^{V}) – 2.5% (total As)
General population (EAT)[d]	0.25%(P1) – 0.72%(P2) – 0.313% (P3)	0.1% (P4)	1.8%	0.2% (total As)
Mean saturation as % of basal value***				
Blood (basal value)[b]	62% (1 µg L^{-1} blood)	42% (90 µg L^{-1} blood)	37% (10 µg L^{-1} blood)	n.d.
Urine (basal value)[b]	35% (2 µg g^{-1} creatinine)	23% (25 µg g^{-1} creatinine)	n.d.	280% (10 µg g^{-1} creatinine for inorganic As)[e]
Risk category[d]	T; Cat. 1 IARC (human carcinogen)	T + N	T	T + N; Cat. 1 IARC[c]
Toxicity[d]	Renal damage, bone lesions, delayed fetal growth, and reduced fertility	Neurotoxicity (saturnism), hematological toxicity (anemia), congenital anomalies	Neurological damage, kidney failure, digestive tract inflammation	Acute: digestive disorders; chronic: cancers of skin, lung, bladder, and kidney. Skin disorders[j]

PTWI, provisional tolerable weekly intake; PTMI, provisional tolerable monthly intake

[a] Merian et al. (2004); [b] CALIPSO (Leblanc et al. 2006); [c] INRS (2010) Toxicology data sheets; [d] EAT (2004); [e] EFSA (2009); [f] JEFCA (2010a); [g] JEFCA 2010b

*Each sample consists of five sub-samples at most, weighted by main place of purchase main place of supply used by consumers on the Secodip panel. Analyses involved an amount of about 0.6 g per composite sample and replicate analyses were performed on each sample

**Each sample of fresh product analyzed consists of about 1000 g of product, i.e., five sub-samples of 200 g. The origin and distribution of the five sub-samples was determined according to place of purchase selected from data on frequency of purchase in the consumer survey, which were weighted by frequency of consumption and quantities consumed

***Basal value: value found for the 95th percentile of the general French population not occupationally exposed (EAT 2004)

n, number of samples; n.d., not determined; T, toxic; N, dangerous for the environment

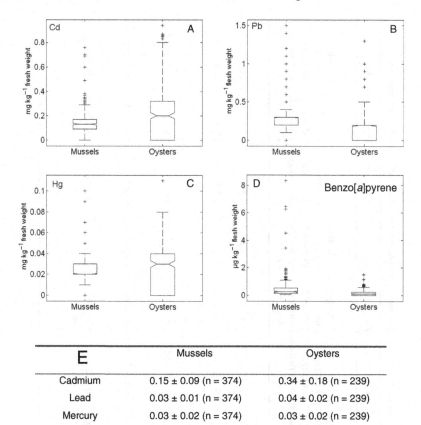

E	Mussels	Oysters
Cadmium	0.15 ± 0.09 (n = 374)	0.34 ± 0.18 (n = 239)
Lead	0.03 ± 0.01 (n = 374)	0.04 ± 0.02 (n = 239)
Mercury	0.03 ± 0.02 (n = 374)	0.03 ± 0.02 (n = 239)
Benzo[a]pyrene	0.56 ± 1.01 (n = 180)	0.27 ± 0.24 (n = 180)

Fig. 1 Distribution of contamination in mussels and oysters in French shellfish-farming areas from 2003 to 2007 (data from Claisse et al. 2006 for 2003–2005; unpublished results from the same authors for 2006–2007 period). (**a**) Cadmium; (**b**) lead; (**c**) mercury in mg kg^{-1} fresh wt, and (**d**) benzo[a]pyrene in μg kg^{-1} fresh wt (**e**) provides values used to construct graphs **a–d**

(*Fucus* spp.) and the environment (Bryan 1976; Bryan et al. 1983). Among other unregulated metals, zinc and magnesium levels are higher in oysters than in mussels (Table 4).

Polonium (^{210}Po) is one of the radionuclides that may have a health impact (exposure threshold 2 millisievert (mSv) yr^{-1}; Table 5). Exposure by ingestion is significant, and annual intake can reach hundreds of microsievert per year in adults (Pradel et al. 2001).

3.2 Organic Contaminants

Bivalve mollusks are exposed to a multitude of persistent or non-persistent organic contaminants belonging to very different chemical families. Tables 6 and 7 give

Table 4 Levels of contamination in the environment and in shellfish sampled from the marketplace for unregulated inorganic contaminants

	Ni	Cr	V	Mn	Cu	Zn	Co	Se	Mg	Mo
Anthropogenic sources	Industry (production of stainless steel, catalysis, etc.)[b]	Industry (anti-corrosion, catalysis, pigments, etc.)[b]	Titanium industry, ports, petro-chemicals[c]	Industry (catalysis, battery manufacture, etc.)[b]	Electrical industry, construction, etc.[b]	Industry (anti-corrosion, coatings, alloys, etc.)[b]	Industry (alloys, pigments, fertilizers, etc.)[b]	Industry (electrical, metallurgy, etc.)[b]	Industry (chemical, alloys, etc.)[b]	Industry (alloys, catalysis, pigments, etc.)[b]
Mean levels in the environment										
Seawater (μg L^{-1})	0.6[a]	0.2[a]	1.9[a]	0.01[a]	0.005–0.05[a]	0.5–5[a]	0.002[a]	0.09[a]	1.3 10^6 [h]	n.d.[h]
Sediments (μg g^{-1} dry wt)	45[a]	60[a]	252[a]	1.2[a]	5–50[a]	50–500[a]	0.045[a]	1.7 10^{-4} a	45[h]	8.10^{-4} a (b)
Mean contamination of shellfish (mg kg^{-1} fresh wt)										
Oysters (min–max) (n = 6)*	0.03–0.17[f]	0.02–0.15[f]	6.3[d]	3.18–7.07[f]	6.90–30.1[f]	111–312[f]	0.01–0.05[f]	0.011[f]	590–957[f]	0.02–0.20[f]
Mussels (min–max) (n = 6)*	0.20–0.53[f]	0.07–0.25[f]	7.3[d]	1.32–3.68[f]	0.89–2.39[f]	8.23–26.7[f]	0.07–0.18[f]	0.011[f]	160–673[f]	0.05–0.51[f]
Recommended nutritional intake per day (d^{-1})				2–3 mg d^{-1} g	0.8–2 mg d^{-1} g	6–19 mg d^{-1} g	0.6 μg d^{-1} g	20–80 μg d^{-1} g	80–420 mg d^{-1} g	30–50 μg d^{-1} g (a)

Table 4 (continued)

	Ni	Cr	V	Mn	Cu	Zn	Co	Se	Mg	Mo
Anthropogenic sources	Industry (production of stainless steel, catalysis, etc.)[b]	Industry (anti-corrosion, catalysis, pigments, etc.)[b]	Titanium industry, ports, petro-chemicals[c]	Industry (catalysis, battery manufacture, construction, etc.)[b]	Electrical industry, etc.[b]	Industry (anti-corrosion coatings, alloys, etc.)[b]	Industry (alloys, pigments, fertilizers, etc.)[b]	Industry (electrical, metallurgy, etc.)[b]	Industry (chemical, alloys, etc.)[b]	Industry (alloys, catalysis, pigments, etc.)[b]
Intake not to exceed	n.d.	n.d.	$100 \ \mu g \ d^{-1}$ [g]	$4.2{-}10 \ mg \ d^{-1}$ [g]	n.d.	$15{-}40 \ mg \ d^{-1}$ [g]	$200 \ \mu g \ d^{-1}$ [g]	$150 \ \mu g \ d^{-1}$ [g]	$750 \ mg \ d^{-1}$ [g]	$350 \ \mu g \ d^{-1}$ [g]
Intake from shellfish ingestion in adult men	$0.76 \ \mu g \ d^{-1}$ [f]	$0.23 \ \mu g \ d^{-1}$ [f]	n.d.	$0.01 \ mg \ d^{-1}$ [f]	$0.02 \ mg \ d^{-1}$ [f]	$0.11 \ mg \ d^{-1}$ [f]	$0.12 \ \mu g \ d^{-1}$ [f]	$0.03 \ \mu g \ d^{-1}$ [f]	$1.2 \ mg \ d^{-1}$ [f]	$0.33 \ \mu g \ d^{-1}$ [f]
Risk category	Xn + T (monoxide)[b]	T + N; Cr(VI) Cat. 1 IARC (human carcinogen)[b]	Xn (divanadium pentoxide); combustible (vanadium trioxide)[e]	Xn (Mn dioxide)[b]	Xn[b]	C (chloride); Xi (sulfate); T (chromate) + N[b]	T + N (Co sulfate); Xn (cobalt)[b]	T + N[b]	Xi (Mg chloride)[b]	Xi[b]

Table 4 (continued)

	Ni	Cr	V	Mn	Cu	Zn	Co	Se	Mg	Mo
Anthropogenic sources	Industry (production of stainless steel, catalysis, etc.)[b]	Industry (anti-corrosion, catalysis, pigments, etc.)[b]	Titanium industry, ports, petro-chemicals[c]	Industry (catalysis, battery manufacture, construction, etc.)[b]	Electrical industry, etc.[b]	Industry (anti-corrosion coatings, alloys, etc.)[b]	Industry (alloys, pigments, fertilizers, etc.)[b]	Industry (electrical, metallurgy, etc.)[b]	Industry (chemical, alloys, etc.)[b]	Industry (alloys, catalysis, pigments, etc.)[b]
Toxicity	Digestive disorders; probable carcinogenic effect[b]	Cr(VI): digestive disorders; kidney failure[b]	Severe systemic poisoning symptoms and death (divanadium pentoxide); headaches, vomiting (vanadium trioxide)[e]	Chronic: nervous and respiratory disorders[b]	Chronic: hepatitis; neurological disorders[b]	Digestive disorders[b]	Irritative respiratory syndrome[b]	Digestive disorders; neurological signs[b]	Muscular tetany, digestive disorders[f]	Diarrhea, anemia, erythrocytic immaturity, uricemia[f]

[a]Merian et al. (2004); [b]INRS (2010) Toxicology data sheets; [c]Saavedra et al. (2004); [d]Roux et al. (2001); [e]International chemical safety sheets (2010); [f]EAT (2004); [g]AFSSA (2008b); [h]OSPAR (2008)

*The shellfish contamination data were obtained from an individual composite sample of five sub-samples at most, weighted by main place of purchase used by consumers on the Secodip panel. Analyses involved an amount of about 0.6 g per composite sample and replicate analyses were performed on each sample n.d., not determined; T, toxic; N, dangerous for environment; Xn, noxious; C, corrosive; Xi, irritant; [a]estimated adult requirement – no DRI value; [b]estimated concentration

Table 5 Radionuclides in the environment and in shellfish sampled from the marketplace

	Radionuclides: ^{99}Tc, ^{129}I, ^{226}Ra, ^{210}Po, ^{238}U, ^{239}Pu, ^{240}Pu, and ^{241}Am
Anthropogenic sources	Nuclear industry; fertilizer manufacture[a]; mining[c]
Mean levels in the environment	
Seawater (μg L^{-1})	^{137}Cs 0.002–0.500 Bq L^{-1} [a]
	^{99}Tc 0.350 Bq L^{-1} [a]
	^{210}Po 1–5 Bq m^{-3} [c]
Sediments (μg g^{-1} dry wt)	^{210}Po 9–125 Bq kg^{-1} [c]
Mean contamination of shellfish	
Mussels (min–max)	^{210}Po 150–600 Bq kg^{-1} dry wt[c]
Cockles (min–max)	^{210}Po 80–1200 Bq kg^{-1} dry wt[c]
Mollusks (mean)	^{210}Po 15 Bq kg^{-1} dry wt[b]
Intake not to exceed	Men 2 mSv yr^{-1} (probable maximum individual dose)[a]
Maximum estimated intake from shellfish ingestion, adult men	^{210}Po 160 μSv yr^{-1} [b]
Risk category	Radiological and chemical risk[a]
Toxicity	Irradiation, contamination, cancers

[a]OSPAR (2007); [b]Pradel et al. (2001); [c]IRSN (2010)

a summary description of the main data available in the literature on pollutants identified in water, sediments, and bivalve mollusks (Leblanc et al. 2006; OSPAR 2008) and include information on toxicity and risk category.

In regard to regulated organic contaminants (Table 6), PCBs and dioxins (PCDD/Fs) are found at levels far below the regulatory thresholds (8 pg g^{-1} of DL-PCBs + dioxins) in oysters (<0.6 pg g^{-1}), mussels (<0.6 pg g^{-1}), and king scallops (<0.4 pg g^{-1}). The benzo[a]pyrene sanitary threshold is exceeded in neither marketed mussels (Table 6) nor those that are farm sourced (Fig. 1d and e). Some data on contamination of shellfish flesh are also available for unregulated organic contaminants (Table 7). Of about 100 existing organostannic compounds, mono-, di-, and tributyltin (MBT, DBT, and TBT) and mono-, di-, and triphenyltin (MPT, DPT, and TPT) are most frequently found in fishery products. Octyltins are not detected in fishery products. Based on the available data, results of two recent studies were that exposure to organotins through seafood does not seem to present a risk for the adult consumer (AFSSA 2006; Guérin et al. 2007). There are other relevant contaminating organic compounds, but very few data are available for them:

- synthetic musks, nitro-musks, and polycyclic musks from the perfume industry;
- octylphenol ethoxylates (OPEs) and nonylphenol ethoxylates (NPEs), from industrial cleaning, maintenance of public places, and processing of leather and textiles;

Table 6 Levels of contamination in the environment and shellfish flesh sampled from the marketplace for regulated organic contaminants under Regulation (EC) No. 1881/2006 (EC 2006c)

Contaminants	PCBs	Dioxins and furanes	PAHs
Sources	Industrial products: transformer and condenser oils[b]	Incineration, metallurgy processes[b]	Constituents of crude oil, incineration, and incomplete burning of organic matter: wood, coal, heating oil[b]. Oil production. Offshore activities[b]. Coal tar coatings, exhaust gases.[b]. Wildfire, volcanic eruptions[b]
	Paint plastifiers and plastics, sealants[b]	Use of active chlorine for bleaching paper pulp[b] Internal combustion engines, wildfire, wood burning[e]	
	Current reservoirs: soil, sediments, rubbish dumps/landfills, old infrastructures[b] Remobilization of old sediments (dredging)[b] Rivers, atmosphere, and ocean currents[b], professional or recreational nautical activities[c]		
Mean levels in the environment			
Seawater (ng L^{-1})	0.001[b]	n.d.[e]	Benzo[a]pyrene 0.001–0.005[b] Fluoranthene 0.036–0.285[b] Benzo[b+k]fluoranthene 0.001–0.017[b] Pyrene 0.011–0.053[b] Total PAHs <0.0001–8500[b]
Sediments (µg kg^{-1} dry wt)	Congeneric PCBs (28/52/101/138/153/180) <0.010–0.116[b]	0.020[e]	Benzo[a]pyrene 0.2–112[b] Fluoranthene 0.72–160[b] Benzo[b+k]fluoranthene 1.1–434[b] Pyrene 0.6–128[b] Total estuarine PAHs 200–6000[b]

Table 6 (continued)

Contaminants	PCBs	Dioxins and furanes	PAHs
Mean contamination of shellfish pg g^{-1} fresh wt	$\sum$ **DL-PCBs**	$\sum$**PCDD/F)**	$\sum$**PAH**
Oysters	0.324[a]	0.272[a]	
Mussels	0.334[a]	0.228[a]	39.0–337[e]
Scallops	0.193[a]	0.199[a]	
Regulatory thresholds[d]	$\sum$(PCDD/F + dl-PCB) 8.0 pg g^{-1} d	$\sum$PCDD/F 4.0 pg g^{-1} d	Benzo[a]pyrene 10.0 pg g^{-1} d
TDI (ng kg^{-1} bwt d^{-1},)	$\sum$PCB 20 (Aroclor eq.)[c]	0.001–0.004[b]	
PTMI (pg kg^{-1} bwt month^{-1})	$\sum$PCBi 10[g]	$\sum$(PCDD/F + $\sum$ dl-PCBs) 70[f]	
Daily intake from food[c]		$\sum$(PCDD/F + dl-PCB): 1.8 pg WHO-TEQ kg^{-1} bwt d^{-1}	$\sum$(6 PAH): 1.4 ng WHO-TEQ/kg bwt/d
Toxicity	Endocrine disruptor	Chloracne[c]	Endocrine disruptor. Benzo[a]pyrene: carcinogen[c]. Less bioaccumulative and
Ecotoxicity	Neurotoxic, immunotoxic[b]	Immunodepressor[c] Carcinogen (2,3,7,8-TCDD)[c]	biomagnifying than organochlorines. Slow metabolization in mussels/fish[b]
Status	Main applications banned in France (1987). Total end to use in 2010[b]	Two decrees in 2002 on waste incineration – limit value 0.1 ng TEQ m^{-3} [e]	Decrees in 1999 limiting PAH emissions to 0.1 mg Nm^{-3} for boilers and engines[e]

[a]CALIPSO study (Leblanc et al. 2006); [b]OSPAR 2008; [c]AFSSA 2008b; [d]Regulation (EC) 1881/2006 (EC 2006c); [e]INERIS (2010) Toxicology data sheets; [f]JECFA (2001); [g]AFSSA (2005)
PCBi: sum of PCBs 28, 52, 101,138, 153, and 180 (AFSSA 2005)

Table 7 Levels of contamination in shellfish sampled from the marketplace for some unregulated organic contaminants

	TBT organostannic compounds	HCB	Dieldrin	DDT/DDE/DDD Total DDT	Lindane α-, β-, γ-HCH	Toxaphene	Triazines atrazine, simazine	Dichlorvos	Brominated flame retardants, polybromodiphenyl ethers (PBDEs)	Chlorinated paraffins
Sources	Agricultural processes, anti-fouling agents, fungicide agents,[b] waste tips[b]. Current reservoirs: soil, sediments, public waste tips and old infrastructures, remobilization of old sediments (dredging)[b]. Direct inputs from rivers, atmosphere, and ocean currents,[b] professional and recreational nautical activities[c]		Environmental stocks[b]		Agriculture, wood treatment, veterinary medicine, domestic use[d]. Control of crop pests, cattle parasites, and commensal insects[d]	Pesticide, complex mixtures of organochlorines[b]	Pesticide[b]	Used on salmon farms: parasiticide against sea louse, insecticide – acaricide[b]	Automobile and aeronautic padding, textile and polymer additives[b], water purification, photography	Plastifiers, additives for metal working fluids, flame retardants, leather industry[b]
Mean levels in the environment										
Seawater (ng L^{-1})	0.6[c]	≈ 1[b]	0.005–0.05[b]	0.005–0.05[b]	0.0005–0.005[b]		< 2–42[b]		DeBDE 1 – 1700[b]	
Sediments (μg kg^{-1} dry wt)	0.001[c]	0.040–0.070[b]	0.0005–0.005[b]	0.0005–0.005[b]						
Ecotoxicological evaluation criteria* (mg/kg fresh wt)						10[b]			10[b]	

Table 7 (continued)

	TBT organostannic compounds	HCB	Dieldrin	DDT/DDE/DDD Total DDT	Lindane α-, β-, γ-HCH	Toxaphene	Triazines atrazine, simazine	Dichlorvos	Brominated flame retardants, polybromo-modiphenyl ethers (PBDEs)	Chlorinated paraffins
Mean contamination of shellfish										
Mussel (μg kg⁻¹ dry wt)	1.1^g			DDE 5–50^b	≈ 1^b	n.d.	n.d.	n.d.	n.d.	n.d.
Mussel (μg kg⁻¹ fresh wt)										
ADI (ng kg⁻¹ bwt d⁻¹) or other TRV	$\sum$(organoSn) 160^b 250^f		Aldrin + dieldrin 0.0001mg kg⁻¹ bwt d⁻¹	0.005 mg kg⁻¹ bwt d⁻¹			Atrazine: 500 ng/kg bwt d⁻¹ Simazine: 520 ng kg⁻¹ bwt d⁻¹	80 ng kg⁻¹ bwt d⁻¹e DL$_{50}$ rat 17–80 mg kg⁻¹ bwt^d DL$_{50}$ mouse 61–135 mg kg⁻¹ bwt^d DL$_{50}$ rabbit 10–12. mg kg⁻¹ bwt^d DL$_{50}$ dog 100 mg kg⁻¹ bwt^d	No TRV exists^a LOAEL octaBDE: 8 mg kg⁻¹ bwt d⁻¹ LOAEL pentaBDE: 72 mg kg⁻¹ bwt d⁻¹	
Daily intake from shellfish ingestion, adult men										
(ng ind⁻¹ d⁻¹)	Ratio (ng ind⁻¹ d⁻¹)/(ADI x 60) = 0.0034^f	n.d.	n.d.	n.d.	n.d.				Fish and seafood: 85^b 150^a	
Risk category	T. N^a				R23/24/25. R36/38^d			R24/25^d		

Table 7 (continued)

	TBT organostannic compounds	HCB	Dieldrin	DDT/DDE/DDD Total DDT	Lindane α-, β-, γ-HCH	Toxaphene	Triazines atrazine, simazine	Dichlorvos	Brominated flame retardants, polybromodiphenyl ethers (PBDEs)	Chlorinated paraffins
Toxicity/ecotoxicity	TBT: Endocrine disrupter[a] TPT: toxic for reproduction and development[a] DBT. TBT. TPT: immunotoxic[a]				Neurological disorders[d]			Acetylcholinesterase inhibitor Mutagenic Carcinogenic Reprotoxic[d]	Endocrine disrupter Neurotoxic, Potentially carcinogenic[b]	
Physico-chemical properties and phenomena determining fate of contaminants					Poorly hydrosoluble. Highly soluble in organic solvents[d]			Poorly hydrosoluble. Soluble in organic solvents[d]	Highly lipophilic. Poorly hydrosoluble Adsorbs strongly to sediments[b]	
Status	Total ban since 1 Jan 2008[c]	Banned[b]		Banned[b]		Not used in OSPAR area[b]	Banned in France[b] Limited uses[b]			End of use for short-chain paraffins scheduled[b]

TBT, tributyltin; TPT, triphenyltin; DBT, dibutyltin; HCB, hexachlorobenzene; T, toxic; N, dangerous for environment; DDT, dichlorodiphenyltrichloroethane; DDE, dichlorodiphenyldichloroethylene; DDD, dichlorodiphenyldichloroethane; HCH, hexachlorocyclohexane; n.d., not determined; R23/24/25, toxic by inhalation, skin contact, ingestion; R36/38, eye and skin irritant; R24/25, toxic by inhalation, skin contact, ingestion

[a]CALIPSO (Leblanc et al. 2006); [b]OSPAR (2008); [c]AFSSA (2006); [d]INRS (2010) Toxicology data sheets; [e]INERIS (2010); [f]EFSA (2004b); [g]Guérin et al. (2007)

- hydrocarbons, particularly toluene, ethyl benzene, xylene (BTEX), and phenols, from the offshore oil industry via sludge and drill cuttings, process water, and accidental spills or illegal discharges;
- substances on the list of 33 priority substances in Annex X of Directive (EC) No. 105/2008 (EC 2008b), especially alachlorine, chloroalkanes, chlorfenvinphos, chlorpyrifos, di(2-ethylhexyl)phthalate (DEHP), diuron, endosulfan, hexachlorobutadiene, isoproturon, pentachlorobenzene, pentachlorophenol, trichlorobenzene, and trifluralin;
- emerging contaminants including pharmaceuticals, hormones, and endocrine-disrupting compounds also present in aqueous environment (Richardson and Ternes 2005).

3.3 Accumulation of Contaminants in Mollusks and Factors of Variation

Shellfish are filter feeders that concentrate contaminants, and also have the ability to detoxify contaminants by themselves. The balance between these two processes is not fixed but depends on many factors.

Contamination may be direct (from water) or via food ingestion. Food contamination in filter-feeding mollusks occurs via seston (suspended particulate matter, inert or living). As with inert particles, phytoplankton becomes contaminated by adsorbing chemical compounds onto their cell surfaces; sometimes, these chemicals are absorbed by diffusion into the cells. Food contamination (phytoplankton) generally leads to longer half-lives than does direct water contamination. The longer the duration of contact, the higher the level of contamination and the longer the decontamination. The ratio of organic to inorganic contaminants influences their distribution in organisms and their elimination rate.

3.3.1 Bioconcentration Factors (BCFs)

The concentration factor (CF) or the bioconcentration factor (BCF) is a concept that was introduced by Polikarpov (1960). It is based on a relatively simple concept that a relationship exists between the concentration of a substance in an organism and the concentration of the same substance in the surrounding water. However, CFs are not easy to estimate; to do so, the two concentrations must remain constant. It is difficult to experimentally maintain constant concentrations in water for long periods of time, and in situ water concentrations fluctuate widely. No method for standardizing the estimation of CFs has been proposed. Numerous studies have been carried out to address this problem (Chong and Wang 2001; James et al. 2006; Miramand et al. 1980; Murray et al. 1991; Pruell et al. 1986). CF data for various organic pollutants have been recorded by different agencies (e.g., the International Atomic Energy Agency (IAEA) and the Groupe Radioécologique Nord-Cotentin) and have been published (Amiard-Triquet and Amiard 1980). CF values vary widely among different animal types and the resultant bioaccumulation values are influenced by many abiotic and biotic factors.

The best estimations of CFs are those that are determined in experiments that are performed in situ over long periods of time. Since the Water Framework Directive (WFD), Directive (EC) No. 60/2000 (EC 2000), has come into force, water authorities are obliged to assess concentrations of pollutants in total seawater, dissolved concentrations, and amounts in particulates. However, hydrophobic pollutants are essentially adsorbed onto particulates and their concentration is dependent on the concentration of these particulates in water. Such particulate concentrations fluctuate widely in space and time, so direct measurements in water were abandoned more than 20 years ago, under the French National Monitoring Network (RNO – *Réseau national d'observation*) and the OSPAR convention. The French Research Institute for Exploitation of the Sea (IFREMER – *Institut français de recherche pour l'exploitation de la mer*) considers that, at least for non-hydrophilic substances, the most effective monitoring target for contaminants are media that concentrate these substances: sediments and/or biota and particularly mussels and oysters, the two usual sentinel species. However, to meet the requirements of the WFD, the levels measured in these media must be converted into water concentrations. The tissue concentration in the mollusks is equal to the concentration in the water multiplied by the BCF. It is therefore possible to calculate the water concentrations, if the CF is known. James et al. (2006) provided BCFs for most substances that the EU considers to be priority ones (Table 8).

3.3.2 Seasonal Fluctuations in Contaminant Concentrations

Concentrations of chemical contaminants in bivalve mollusks fluctuate according to the time of year. This was noticed from the start of the RNO monitoring program in the early 1980s (Claisse 1992). The pattern for inorganic compounds is "biological dilution" when bivalves reach sexual maturity; this occurs when the amount of contaminants remain the same, but the organism's body mass increases, and thus metal concentrations fall. This has been observed for cadmium, copper, lead, and zinc in mussels (Amiard et al. 1986) and oysters (Amiard and Berthet 1996). The highest concentrations are recorded in winter and spring and the lowest in summer and autumn, with ratios of up to 1:4 depending on the contaminant and the species (Devier et al. 2005). The reverse pattern is found with lipophilic organic compounds, such as DDT in the oyster *C. virginica*; concentrations increase at sexual maturity when oysters produce lipid-rich gametes (Butler 1973). Oysters also eliminate these pollutants through spawning (release of eggs into the water). With *C. virginica*, the risk to humans is therefore greatest at the moment of sexual maturity.

Because contaminants are monitored only annually, and because of the kinetic behavior of contaminants in mollusks, tracing individual contamination events over short periods of time is not possible. Therefore, the established programs are effective for monitoring chronic contamination, but not for short duration events; such events may thus go unnoticed between any two samplings of the sentinel species.

Table 8 Bioconcentration factors for chemical contaminants in bivalve mollusks

Substance	BCF in mollusks
Anthracene	260 (*Macoma*)
Cadmium	994 (invertebrates)
C10-13 chloroalkanes	40,900 (mussels)
Chlorfenvinphos	255 (*M. galloprovincialis*)
Diethylhexyl phthalate	2,500 (mussels)
Endosulfan	600 (*Mytilus*)
Fluoranthene	10,000 (*Crassostrea*)
Hexachlorobenzene	7,000 (bivalves)
Hexachlorobutadiene	2,000 (*Mytilus*)
Hexachlorocyclohexanes (lindane)	161 (mussels) 240 (*Mytilus*)
Lead	2,279 (mollusks)
Mercury[a]	10^6–10^7
Naphthalene	27–38 (mussels)
Nickel	270 (bivalves)
Nonylphenols	3,000 (mussels)
Octylphenols	634 (calculated)
Pentachlorobenzene	2,000 (bivalves)
Pentachlorophenol	390 (*Mytilus*)
Benzo[a]pyrene	12,000 (*Mytilus*)
TBTs	11,400 (*Crassostrea*)
Trifluralin	2,360 (*Helisoma*)
Aldrin	43,560 (calculated)
Dieldrin	7,760 (calculated)
Endrin	5,250 (calculated)
Isodrin	43,650 (calculated)
Total DDT	45,600 (mollusks)

Source: James et al. (2006)
[a]Bioamplification taken into account

However, alarms may be sounded from accidental discharges as a result of triggering increased mortality at sensitive developmental stages.

3.3.3 Detoxification Mechanisms

Detoxification of Trace Elements

Invertebrates exposed to toxic trace elements respond with two types of detoxification mechanisms (Amiard 1991). The first response is to render the metal insoluble by immobilizing it in the form of a salt. This occurs with silver sulfide in oysters, for example (Martoja et al. 1988). The second response is to induce metallothioneins (MTs), which are capable of detoxifying various trace elements (Amiard et al. 2006). MTs form complexes with the trace elements and render them harmless. Metallothioneins are stored in lysosomes and their concentration is proportional to that of toxic trace elements in the environment, as shown by an experiment with transplanted mussels in the western Mediterranean Sea (Mourgaud et al. 2002). Detoxification mechanisms in invertebrates vary widely from one species to another.

In various oyster species, mobile cells called amebocytes accumulate complexed metal from the blood. In *Ostrea edulis*, some amebocytes accumulate copper, others zinc, or copper and zinc simultaneously. Other oyster species, such as *Ostrea angasi* and *C. gigas*, have only one amebocyte type, which accumulates copper and zinc equally well (George et al. 1984). Some species of mollusks (e.g., oysters and mussels) are capable of regulating the internal concentration (homeostasis) (within certain concentration limits) of certain essential trace elements, such as copper and zinc (Amiard et al. 1987).

The particular physical–chemical form of inorganic contaminants that are stored have consequences for the subsequent transfer of trace elements within trophic networks. The two above-mentioned detoxification processes (insolubilization and metallothionein induction) are very efficient, and species that use them can live in heavily contaminated environments. Such species may accumulate high levels of contaminants in some of their tissues. When these species are consumed, the metal–metallothionein complexes are ingested and digested, releasing the metals into the consumer's body in a manner that favors the assimilation of the metals. Therefore, the levels transferred to and absorbed by the consumer may be high. In contrast, when detoxification occurs by insolubilization, the resultant granules are poorly digested by the consumer or the predator; hence, bioavailability is low.

Detoxification of Organic Pollutants

Some invertebrates are able to biotransform organic pollutants in special organs (e.g., the digestive gland) that render pollutants hydrosoluble, and therefore more easily eliminated (Narbonne and Michel 1997). This metabolic process occurs in two biotransformation stages: (1) phase I, oxidation, and/or (2) phase II, conjugation. Phase I is controlled by P450 cytochromes or flavin monooxygenases. In phase II, conjugation frequently takes place with glutathione and is catalyzed by glutathione *S*-transferase (GST). Occasionally, biotransformation activates a metabolite to a form that is more toxic than the parent molecule. A third detoxification pathway is possible and involves the glycoprotein Pg-170 (phase III). In phase III, organic pollutants are expelled from the cell. This protective elimination mechanism is efficient in mollusks (Bard 2000) and is known as multixenobiotic resistance (Pain and Parant 2003).

The Effect of Shellfish Purification on Chemical Contaminants

In the course of shellfish production, shellfish are purified to reduce the risk of microbiological contamination. The question is whether this microbiological purification helps reduce the amount of any chemical contaminant also present in the shellfish.

Microbiological purification consists of immersing live shellfish in tanks continuously fed clean seawater for a period that is sufficient to eliminate microbiological contaminants and render the shellfish suitable for human consumption. The regulatory definition of "clean seawater" is found in point *h* of Article 2 of Regulation (EC)

852/2004 (EC 2004b). This very vague definition sets goals, without clearly defining the criteria to be fulfilled. The French Directorate for Food (DGAL), therefore, commissioned AFSSA to establish seawater quality criteria suitable for handling fishery products. AFSSA delivered its opinion on 26 July 2007 (AFSSA 2007a). Microbiological purification is required only for shellfish from Class B and C production areas, and the produce from these areas can be harvested, but cannot be directly marketed. The time required for purification varies between two and several days, depending on the system used. In France, the duration for purification is 48 h for Class B shellfish (industry recommendation). The duration of purification may be reduced for some fragile shellfish species (e.g., wedge shells and *Ruditapes* clams); the regulations do not impose a minimum duration.

When kept in large quantities of clean seawater, contaminated marine organisms purify themselves, eliminating the chemical contaminants that they have accumulated in their soft tissues. The measure used to track elimination rate is biological half-life, i.e., the time required for half the amount of a substance to disappear from the organism or the organ.

The kinetics of decontamination depends on not only the difference in initial concentration but also the following factors (Casas and Bacher 2006):

– chemical-specific factors (type(s) of the contaminant(s), level(s) of contamination, variations in contamination over time, contamination pathways (i.e., water, food, or inert particles));
– physiological factors of the organism (growth rate, mass variation over time, type of sexual state maturity, physiological status, differences between species, etc.); and
– environmental factors (temperature, and food quantity and quality).

From the foregoing, it is obvious that the elimination kinetics, the mechanisms of elimination, and quantities of toxicants eliminated will be species dependent. Mussels are capable of eliminating cellular organelles (lysosomes) that were involved in detoxifying various contaminants, whereas oysters retain their lysosomes for life (George et al. 1978). In some species, certain cumulative toxins continue to be accumulated throughout an animals' lifetime.

In Table 9 we provide examples of the chemical half-lives of several contaminants in bivalve mollusks. Although this table is far from exhaustive, it indicates the wide variations in half-life elimination times for various contaminants and species.

The above information disclosed on elimination half-lives of various chemicals indicates that the 48 h immersion time, used to purify microbes from Class B shellfish, is far from sufficient to also remove chemical contaminants (organic and metal). In fact, considerably more research results are needed to achieve reliable estimates of the half-lives in shellfish species of the main contaminants found in the marine environment. These data would be extremely useful in estimating the dissipation times, and therefore the seriousness of accidental chemical pollution or spills. Of course, the key question after such events occur is how soon and under what

Table 9 Example half-lives for chemical contaminants that exist in bivalve mollusks

Species	Chemical contaminant	Biological half-life (days)	Reference
M. edulis	TBT	21–36	Yang et al. (2006)
	TBT	69	Page et al. (1995)
	DBT	115	Page et al. (1995)
	Fluoranthene	30	Pruell et al. (1986)
	benzo[*a*]anthracene	18	Pruell et al. (1986)
	Chrysene	14	Pruell et al. (1986)
	Benzo[*e*]pyrene	14	Pruell et al. (1986)
	Benzo[*a*]pyrene	15	Pruell et al. (1986)
	Indeno[*1,2,3-cd*]pyrene	16	Pruell et al. (1986)
	PCB 28	16	Pruell et al. (1986)
	PCB 101	28	Pruell et al. (1986)
	PCB 128	37	Pruell et al. (1986)
	PCB 153	46	Pruell et al. (1986)
	Zn	76	Bryan (1976)
M. galloprovincialis	Hg	1000	Bryan (1976)
Mya arenaria	TBT	71–94	Yang et al. (2006)
Gafrarium tumidum	Ni	35 ± 7	Hédouin et al. (2007)
Venerupis decussata	TBT	17–38	Gomez-Ariza et al. (1999)
Crassostrea gigas	Cu	11.6–25.1	Han et al. (1993)
	Zn	16.7–30.1	Han et al. (1993)
	Cd	137	Geffard et al. (2002)
	Cu	430	Geffard et al. (2002)
	Hg	44	Bryan (1976)
	Zn	335	Geffard et al. (2002)
	Zn	255	Bryan (1976)
C. virginica	Fluoranthene	26–32	Sericano et al. (1996)
	Pyrene	10–12	Sericano et al. (1996)
	Benzo[*a*]anthracene	13–15	Sericano et al. (1996)
	Chrysene	12–16	Sericano et al. (1996)
	Benzo[*e*]pyrene	12–16	Sericano et al. (1996)
	Benzo[*a*]pyrene	9–10	Sericano et al. (1996)
	Indeno[*1.2.3-cd*]pyrene	10–11	Sericano et al. (1996)
	PCB 26	22	Sericano et al. (1996)
	PCB 118	73–299	Sericano et al. (1996)
	PCB 149	130->365	Sericano et al. (1996)

Table 9 (continued)

Species	Chemical contaminant	Biological half-life (days)	Reference
	PCB 153	51–102	Sericano et al. (1996)
O. edulis	Zn	890	Bryan (1976)
Crassostrea belcheri	Cd	5–16	Lim et al. (1998)
	Cu	5–9	Lim et al. (1998)
	Pb	4–14	Lim et al. (1998)
Crassostrea iredalei	Cd	4	Lim et al. (1998)
	Cu	6	Lim et al. (1998)
	Pb	6	Lim et al. (1998)
Isognomon isognomon	Ni	Infinite	Hédouin et al. (2007)

conditions marketing of exposed shellfish can be resumed. Unfortunately, despite the usefulness of such information for improving shellfish quality, the current regulations do not require that the composition of chemical contaminates in shellfish be considered.

4 Chemical Monitoring in the Environment and in Shellfish

4.1 Environmental Chemical Monitoring Programs

Shellfish are at risk from pollutants primarily because of their environmental exposure. To protect shellfish from chemical contamination, systems have been established to periodically monitor waters of coastal areas for selected contaminants (Apeti et al. 2010; Cantillo 1998; Claisse 1989; Franco et al. 2002; O'Connor 1998). The goal of the OSPAR convention for the protection of the northeast Atlantic marine environment is to reduce pollution. The OSPAR Hazardous Substances Committee listed the substances to be monitored in order of priority, taking into account those that are already prioritized by other regulations, e.g., under the WFD. Under the Mediterranean Action Plan (MAP) of the United Nations Environment Programme (UNEP), the Barcelona Convention for the Protection of the Mediterranean Sea Against Pollution (MED POL) has implemented phase III of the MED POL monitoring program.

European Directive (EC) No. 105/2008 (EC 2008b), which amends Directive (EC) No. 60/2000 (EC 2000) and lays down the environmental quality standards for water, provides for updating the list of priority substances. The updates give the maximum allowable concentration of each substance (set up to avoid serious and irreversible consequences of acute short-term exposure for an ecosystem), as well as the allowable mean annual concentration (to avoid long-term irreversible consequences).

In France, the monitoring of water contamination along the French coast has been performed by the RNO, renamed ROCCH (*Réseau d'observation de la contamination chimique du milieu marin*), in 2008. ROCCH was established by the French Ministry of the Environment in 1974 and is coordinated by IFREMER. Its purpose is to assess levels and trends in chemical contamination along the coast. Until 2007, the RNO monitored only sediments and bivalve mollusks, in which contaminants are concentrated, to meet French obligations under the OSPAR and Barcelona conventions. In addition to sediments and bivalves, ROCCH also monitors the biological effects of contamination by organic forms of tin (which cause imposex; Huet et al. 2003).

4.1.1 Monitoring Contaminants: The RNO Program and Its Successor (ROCCH)

Because of the difficulty in obtaining valid samples suitable for water trace analysis, and the low spatial and temporal representativeness of such samples, RNO monitoring has focused on the matrices that absorb contaminants, i.e., biota and sediments. Therefore, bivalve mollusks (mussels and oysters) are used as quantitative contamination indicators (Claisse 1999).

The concepts of indicator- and sentinel species are widely used in many countries, e.g., Mussel Watch in the USA (Cantillo 1998; Claisse 1989; Goldberg et al. 1983; O'Connor 1998; Sukasem and Tabucanon 1993; Tripp et al. 1992).

In France, testing for chemical contaminants was performed annually in November for all substances and biannually (February and November) for trace elements (Table 10). The interpretation of the analytical results requires consideration of the differences among species in bioaccumulation; for example, the concentration ratios between oysters and mussels are approximately 50 for silver, 2.5 for cadmium, 10 for copper, and 15 for zinc (Claisse et al. 2006).

The RNO results have also sometimes been used for monitoring food safety, together with results from official regulatory controls.

The main achievements of the RNO from 1979 to 2007 included the following:

– establishment of national baseline levels for 9 trace elements, 14 organochlorine chemicals, and 37 PAHs (Table 10);
– identification of reference or control sites for monitoring if

   • natural contaminants are present at representative levels, or
   • synthetic chemicals exist at levels that do not reflect significant inputs, and
   • hotspots exist (particularly contaminated areas; e.g., the Gironde is a hotspot for cadmium and the Seine for PCBs);

– determination of temporal trends for 33 contaminants;
– assembling a bank of stabilized mollusk samples beginning in 1981;

Table 10 The RNO/ROCCH monitoring program for the various conventions and directives [Water Framework Directive (WFD), Oslo and Paris convention (OSPAR), Barcelona convention (MED POL), and for the French Directorate General for Food (DGAL)], with regard to water, biota, and sediment

RNO (1979–2007)

Sampling frequency

Conventions/directives	Water	Biota	Sediment
OSPAR & Barcelona		Annual, in November (at all 80 RNO sites)	Every 10 years (entire French coast)
DGAl		Annual, in February (at all 80 RNO sites)	

RNO contaminants (1979–2007)

Metals	Cadmium (Cd), copper (Cu), mercury (Hg), silver (Ag), chrome (Cr), nickel (Ni), lead (Pb), vanadium (V), zinc (Zn)
Organochlorines	DDT, DDD, DDE, lindane (γ-HCH), α-HCH, polychlorobiphenyls: indicator PCBs (28, 52, 101, 138, 153, 180) and dioxin-like PCBs (105, 118, 156)
Polycyclic hydrocarbons (PAHs)	Naphthalene, mono-, di-, tri-, and tetramethyl naphthalenes, acenaphthylene, acenaphthene, fluorene, mono- and dimethyl fluorenes, phenanthrene, anthracene, mono-, di-, and trimethyl phenanthrenes/anthracenes, fluoranthene, pyrene, mono- and dimethyl pyrenes/fluoranthenes, benzo[*a*]anthracene, triphenylene, chrysene, mono- and dimethyl chrysene, benzofluoranthenes, monomethyl benzofluoranthenes, benzo[*e*]pyrene, benzo[*a*]pyrene, perylene, dibenzo[*a,h*]anthracene, benzo[*g,h,i*]perylene, indeno[*1,2,3-cd*]pyrene, sulfurated heterocycles: dibenzothiophene, mono-, di-, and trimethyl dibenzothiophene, benzonaphthothiophenes, monomethyl benzonaphthothiophenes

Table 10 (continued)

ROCCH (since 2008)			
Sampling frequency			
Conventions/directives	Water	Biota	Sediment
WFD	Monthly for 12 months every 6 years (at all WFD sites)	Annual in November (at 25% of WFD sites)	Every 6 years (at 25% of WFD sites)
OSPAR & Barcelona		Annual in November (at 50% of WFD sites)	Every 6 years (at 50% of WFD sites)
DGAL		Annual in February, Cd, Hg, Pb (on 131 sites)	

ROCCH contaminants (WFD + OSPAR + DGAL)

Metals	Cadmium (Cd), mercury (Hg), nickel (Ni), lead (Pb)
Organic contaminants	Polychlorobiphenyls: indicator PCBs (28, 52, 101, 138, 153, 180) and dioxin-like PCBs (105, 118, 156)
	Alachlor, anthracene, atrazine, benzene, pentabromodiphenyl ether, octabromodiphenylether, decabromodiphenylether, C10–13 chloroalkanes, chlorfenvinphos, chlorpyrifos, 1,2-dichloroethane, dichloromethane, di (2-ethylhexyl)phthalate (DEHP), diuron, endosulfan (family), fluoranthene, hexachlorobenzene, hexachlorobutadiene, hexachlorocyclohexane (alpha, beta, delta), lindane, isoproturon, naphthalene, nonylphenols, 4-n-nonylphenol, $para$-nonylphenols, octylphenol, $para$-$tert$-octylphenol, pentachlorobenzene, pentachlorophenol, benzo[a]pyrene, benzo[b]fluoranthene, benzo[g,h,i]perylene, benzo[k]fluoranthene, indeno[$1,2,3$-cd]pyrene, simazine, tributyltin, trichlorobenzene, 1,2,4-trichlorobenzene, trichloromethane (chloroform), trifluralin, aldrin, carbon tetrachloride, total DDT, p,p'- DDT, dieldrin, endrin, perchloroethylene (tetrachloroethylene), trichloroethylene, isodrin

- organization and management of national and international collaborations through -European conventions and international programs previously cited at the beginning of Section 4.1; and
- implementation of data quality management, which is a driver for achieving the "state of the art" in marine environmental chemical analyses.

Although the RNO was designed for environmental monitoring purposes, it has also performed annual monitoring for food safety purposes to classify the shellfish-farming areas and has conducted discrete site-specific studies.

In 2008, IFREMER established ROCCH (formerly RNO) for the French Ministry of the Environment, although ROCCH is partly financed by water authorities. The main purpose of ROCCH is to address the chemical monitoring needs of the WFD, and the OSPAR and Barcelona international conventions. ROCCH, contrary to RNO, performs chemical monitoring of WFD substances directly in the water, but to the detriment of monitoring shellfish. In particular, the February surveys of shellfish have been discontinued. However, as an annual peak in shellfish contamination was regularly observed, this change may be prejudicial for food safety monitoring, so since 2008, DGAL has financed a February monitoring survey. The number of sampling points has been increased by 60% for this February survey to improve coverage of the shellfish-farming areas. Similarly, the number of taxa monitored has been increased to also address farmed species. Analytical results of the monitoring are published no more than 3 months after the sampling, compared to 10 months post-monitoring under the RNO system.

Up to the present, food safety monitoring has been applied to only three trace elements. However, starting in 2011, DGAL and IFREMER will initiate monitoring for dioxins, DL-PCBs, and benzo[a]pyrene, to comply with Regulation (EC) 1881/2006 (EC 2006c) and to follow the recommendations published in AFSSA's opinion of 21 March 2008 (AFSSA 2008a).

The monitoring work undertaken by RNO and ROCCH are described in Table 10, in the context of the various conventions and directives.

4.1.2 Examples of Contaminant Testing

In this section, a coastal lagoon (Arcachon Bay, *Bassin d'Arcachon* in French) and an estuary (Bay of Seine) have been taken as examples:

Bassin d'Arcachon

The mean concentrations of lead, cadmium, mercury, and other contaminants detected in Bassin d'Arcachon are shown in Table 11. The mean concentrations recorded by the RNO in oysters from Bassin d'Arcachon are 0.18 ± 0.04 mg kg^{-1} fresh wt for lead, 0.23 ± 0.09 mg kg^{-1} for cadmium, and 0.03 ± 0.01 mg kg^{-1} for mercury. These figures are well below the regulatory limits (Table 2). High concentrations of copper are found in oysters (24.51 ± 9.69 mg kg^{-1} fresh wt flesh).

Table 11 Concentrations of certain contaminants (fresh wt)* observed in oysters from Arcachon Bay (RNO survey Feb 2000–Nov 2005) and in mussels (Devier et al. 2005)

Contaminant	Oysters (mean ± s.d.) (*n*)	Mussels (min–max of means depending on site) (*n*)
Inorganic (mg kg^{-1} fresh wt)		
Cadmium	0.23 ± 0.09 (54)	0.14– 0.18 (84)
Lead	0.18 ± 0.04 (54)	0.25–0.31 (84)
Mercury	0.03 ± 0.01 1 (54)	n.d.
Arsenic	n.d.	2.5–2.9 (84)
Nickel	0.21 ± 0.04 (18)	0.20–0.25 (84)
Chrome	0.17 ± 0.08 (42)	0.23–0.34 (84)
Vanadium	0.33 ± 0.12 (18)	n.d.
Copper	24.51 ± 9.62 (54)	1.1–4.1 (84)
Zinc	372 ± 112 (54)	28–42 (84)
Selenium	n.d.	1.6–2.2 (84)
Silver	0.79 ± 0.33 (18)	n.d.
Organic (pg g^{-1} fresh wt)		
Organostannics (amount in Sn)	n.d.	7.2–394 10^3
PCBs (sum of six congeners)	5.2 10^3 ± 3.6 10^3 (21)	5.4–7.0 10^3
PAHs**	40 10^3 ± 11 10^3 (15)	13.3–262 10^3
DDT/DDE/DDD (sum of the three)	2.3 10^3 ± 1.3 10^3 (24)	n.d.
Lindane (α-,γ-HCH) (sum of the two)	0.23 10^3 ± 0.10 10^3 (24)	n.d.

*Fresh weight obtained by multiplying dry weight value by 0.18; n.d., not determined
n, number of samples
**15 PAHS identified as having priority by the EPA

The concentrations have risen over the past 20 years, probably because copper has replaced the TBTs in anti-fouling paints (Claisse and Alzieu 1993).

In regard to the TBTs, mussels transplanted to oyster farms have revealed concentrations of approximately 30 μg kg^{-1} dry wt and showed increases in July and August (Devier et al. 2005). No trace of TBTs has been detected in the water. However, in mussels transplanted to harbor areas, concentrations of 800–2400 μg kg^{-1} dry wt have been recorded, with peaks occurring between April and September. Devier et al. (2005) attribute this increase to spring and summer nautical activities. TBT concentrations measured in the surface waters of Arcachon harbor range between 2 and 7 ng L^{-1} (samples taken from May to August); the corresponding BCF values range from 2.8 × 10^5 to over 1.3 × 10^6. These are the highest BCF values recorded in the literature for mussels (*Mytilus* sp.). TBT levels of 400 μg Sn kg^{-1} dry wt, measured in sediments, are responsible for the high contamination levels found in mussels and result from sediment resuspension (Devier et al. 2005). The observed speed of TBT bioaccumulation is high and is consistent with data in the literature (stabilization after 25 days). Devier et al. (2005) concluded that Arcachon harbor is severely contaminated with organotins because of their persistence in sediments from use as an anti-fouling treatment for boats; the organotins continue as significant contaminants several years after their use has been banned. The concentrations recorded in mussels transplanted to the harbor highlight the role this hotspot plays in local contamination and the hazard it represents for the entire Arcachon Bay. These data confirm the work of Auby and Maurer (2004), who revealed TBT levels (between 1997 and 2003) in Arcachon Bay waters near the harbor service station that ranged from 5.7 to 21.9 ng L^{-1}. The toxic effects on plankton and mollusks associated with these TBT concentrations in seawater have been recorded by Alzieu et al. (1991) and Michel and Averty (1999). They reported that even for a TBT concentration in seawater of less than 1 ng L^{-1}, the females of some gastropods may develop male sexual characteristics (imposex). At concentrations exceeding 1 ng L^{-1}, diatom growth and zooplankton reproduction are restricted; above 2 ng L^{-1}, oyster shells show calcification anomalies, and above 20 ng L^{-1}, reproductive anomalies are observed in bivalves.

High levels of PAHs were measured in mussels transplanted in Arcachon harbor, with peaks occurring in May–June and August (the annual means at this site range from 1.45×10^6 to 1.62×10^6 pg g^{-1} dry wt, depending on the specific PAH, with a maximum of 2.7×10^6 pg g^{-1} dry wt) (Devier et al. 2005).

Regarding indicator PCBs (sum of PCBs 28, 52, 101, 118, 138, 153, and 180), the levels measured in mussels are low (annual means of 5.4 and 7×10^3 pg g^{-1} wet wt). Concentrations in oysters are similar, with $5.2 \pm 3.6 \times 10^3$ pg g^{-1} wet wt.

Twenty-one pesticidal and biocidal active substances have been detected in the waters of the Arcachon Bay during the summertime from 1999 to 2003, at concentrations ranging from a few nanograms per liter to several hundred nanograms per liter. Most of these substances are herbicides, including some that are now banned (Auby and Maurer 2004). According to Auby and Maurer (2004), the presence of these substances may impact the development of the small phytoplankton on which oyster larvae feed, but probably do not affect oyster larval development.

The studies of Auby and Maurer (2004) and Devier et al. (2005) thus emphasize the need to monitor TBT and PAH contamination levels in shellfish-farming areas of Arcachon Bay. Doing so will ensure that TBT and PAH pollution does not migrate from the harbor to the oyster- and mussel-farming areas.

The need to monitor TBT and PAH contamination levels in shellfish-farming areas, as observed at Arcachon Bay, can be extended for the entire French coast, since organostannic and PAH compounds are present in similarly semi-enclosed waters elsewhere along the coast. The highest concentrations of TBTs and their degradation products are recorded in harbor areas, e.g., Brest (1.5 mg kg^{-1} of TBT) and Lorient (0.44 mg kg^{-1}) on the Atlantic coast and Gulf of Fos (1.1 mg kg^{-1}), Toulon (4.1 mg kg^{-1}), and Gulf of Saint-Tropez (1.55 mg kg^{-1}) (Averty et al. 2005) on the Mediterranean coast. Relatively high levels of TBT and PAH are also found in other coastal areas such as the Seine estuary, the Basque coast, and Thau Lagoon.

Bay of Seine

A study of metal contamination of the main marketed species in Bay of Seine was conducted in 2000. The aim was to assess levels of contamination by lead,

Table 12 Concentrations of certain contaminants (mg kg^{-1} fresh wt) observed in mussels from the Bay of Seine (RNO survey from 2003 to 2007)

Contaminant	Mean ± s.d.	Sample size
Cadmium	0.23 ± 0.09	48
Lead	0.49 ± 0.26	48
Mercury	0.04 ± 0.02	48
Benzo[a]pyrene	3.01 ± 4.10 10^{-3}	24

mercury, cadmium, chromium, and silver in five commercial species of interest: whelk, king scallop, plaice/sole, cod, and rock salmon. The study (Chiffoleau et al. 2002) shows that whelks were heavily contaminated with cadmium – above the French regulatory limit in very large specimens (over 70 mm). Based on this finding, a local decree was issued in July 2002, classifying whelks of over 70 mm as "Class D" (French classification grade designating that harvest is prohibited) and whelks of less than 70 mm as "provisional Class A" throughout Bay of Seine and the coasts of Seine Maritime district. In 2002 and 2003, whelk sampling was intensified, particularly for small specimens, to determine the size, on average, above which the 2 mg Cd kg^{-1} wet wt threshold (French decree of 21 May 1999) would be exceeded.

The mean concentrations of cadmium, mercury, lead, and benzo[a]pyrene in mussels are given in Table 12. The concentrations of these four contaminants are below the regulatory limits (Table 2).

4.1.3 Active Environmental Biomonitoring: A Promising Procedure for the Future

Researchers have been conducting active biomonitoring using various shellfish species for several years. For example, the study of Devier et al. (2005) used transplantation experiments. Active biomonitoring has a number of advantages over conventional monitoring (Andral et al. 2004). The transplanted shellfish have a known history, their exposure time is controlled, the citing of the station is chosen independently of bathymetry, and each specimen's position in the water column is controlled. Measurements are optimized, because samples are more homogeneous owing to the selection of specimens for the experiment (parental origin, size, age, healthy site of origin, etc.). There are some constraints, such as complicated logistics and data interpretation that depends on the trophic and physico-chemical variability of the destination site; additional biometric parameters must therefore be measured. The abundant literature in this field (Berthet 2008; De Kock and Kramer 1994; Mourgaud et al. 2002) provides transplantation protocols that include the time required to establish equilibrium with the new environment, the initial stress, and the trophic factors of the destination site.

Transplantation is a promising procedure for the future because of numerous benefits already cited; nevertheless, one aspect thus far neglected is the possibility of theft by ill-intentioned people.

4.2 Chemical Monitoring for Marketed Shellfish

Those who produce and/or market bivalve mollusks are subject to self-inspection and mandatory product traceability to provide information on product quality, including information on content of chemical contaminants and shellfish mortality. For marketed shellfish, the public health authorities responsible for official controls must follow the provisions of the Annex II of Regulation (EC) 854/2004 (EC 2004c). The French Directorate General for Food (DGAL – *direction générale de l'alimentation*) is in charge of these controls and has drawn up annual monitoring programs since 1998 to assess the contamination levels of marketed shellfish.

4.2.1 Self-inspection

Self-inspection is a key tool for shellfish operators to optimize their effectiveness in meeting the requirements of the Hygiene Package. In addition, self-inspection during production, transportation, purification, maturing, and finishing also ensures the food safety of shellfish when they reach the consumer. Self-inspection is carried out for microbiological and chemical contaminants, both in the water and in the shellfish. Sampling is performed by third-party professionals who send their samples to a laboratory of their choice.

4.2.2 Monitoring and Management of Shellfish Mortality

Operators must report each event of mortality that exceeds 20% of individuals within a 15-day period to the responsible authority. IFREMER then conducts a survey to determine the cause of the mortality and whether it has an environmental, a microbiological (often involving *Vibrio*, viruses, fungi, or parasites), or a zootechnical origin. For animal health reasons, IFREMER produces periodic reports on national and regional oyster mortality, through the mollusc pathology network (REPAMO – *Réseau de pathologie des Mollusques*) and other organizations. Mortality occurs in patches within an area and generally affects only one species. It is thought to be multifactorial (Oyster summer mortality program, i.e., MOREST – *mortalité estivale d'huîtres* – and REPAMO) and involve oyster physiology, environmental factors (it does not occur below a temperature of 19°C), and/or aggravating factors (viruses and bacteria) (Samain and McCombie 2008). According to Gagnaire et al. (2006), pesticides may be among the triggering factors.

The epidemiological aspect of these die-offs and the zootechnical and environmental context provide guidelines for diagnosis. For example, if several species are affected simultaneously, an environmental or a toxic origin will be strongly suspected. Blooms of *Gymnodinium*, stress and anoxia are known to cause die-offs. However, it is difficult to precisely identify causes, because operators sometimes take their samples at intervals of 2 weeks or more (e.g., where concessions are accessible only during low spring tides). These mortality events also require dealing with decomposing shellfish, which can affect the microbiological quality of the water in a confined environment. Summer mortality of Pacific oysters (*C. gigas*) on the French coast is regularly reported but has not endangered this species, which was considered to be invasive up until 3 years ago. Recurrent seasonal mortality has also been reported in *Ruditapes* clams and cockles, but not at the same time of year (in spring for *Ruditapes* clams, after stormy episodes for cockles). In 2008 and 2009, there was high mortality among Pacific oysters in France. Laboratory experiments have shown that certain pollutants can affect the genetic, immunity, and trophic characteristics of oysters; in 2009, the combined presence of the OsHV-1 virus and the bacterium *Vibrio splendidus* seems to have played a major part in the mortality incident (Sauvage et al. 2009).

No oyster pathogen is known to also be pathogenic for humans. In some cases of abnormal mortality in marine species (e.g., several species suddenly, simultaneously, and massively affected), a more thorough toxicological investigation may be undertaken to test for pesticides or biocides.

4.2.3 Monitoring Program for Chemical Contaminants in Marketed Shellfish

Two offices of the DGAL are involved in monitoring chemicals in shellfish: (1) the Office for the Quality and Safety of Food Products from Fresh and Marine Waters (BQSPMED – *Bureau de la qualité sanitaire des produits de la mer et d'eau douce*), responsible for monitoring chemical contaminants in bivalve mollusks, and (2) the Office of Food and Biotechnology Regulations (BRAB – *Bureau de la réglementation alimentaire et des biotechnologies*), responsible for the EU dioxin monitoring program. The International Health and Safety Coordination Mission (MCSI – *Mission de coordination sanitaire internationale*, part of DGAL) is also involved by sampling imports. The screened chemical contaminants are trace elements (lead, cadmium, and mercury), indicator PCBs (seven congeners: 28, 52, 101, 118, 138, 153, and 180), and PAHs (15 since 2006). In earlier monitoring and control programs (1998–2002), pesticides and antibiotics (EC 2000) were tested. The number of bivalve mollusk samples to be tested each year, under the chemical contaminants monitoring program, is 400 altogether (all species and all chemical contaminants); this number includes farmed shellfish (oyster, mussel, cockle, and *Ruditapes* clams) and wild populations of pectinids fished in French waters.

The local veterinary authorities (DDSV – *Direction Départementale des Services Vétérinaires*) inform DGAL of positive results without delay. DGAL transfer this information to local Maritime Affairs Authorities (DDAM – *Direction Départementale des Affaires Maritimes*) and to IFREMER. An investigation is then carried out to identify the contamination source and any corrective measures that are required.

In Table 13, the initial screening gave some non-compliance data for cadmium; a second more refined analysis of the same samples was performed by the French National Reference Laboratory (NRL) to address analytical uncertainties, as provided for by Regulation (EC) No. 333/2007 (EC 2007). In this second analysis, only four of the eight samples analyzed were considered to have exceeded the maximum level of 1 mg kg^{-1} fresh wt. In view of the results from 2005, DGAL conducted a control study that had five samplings in March 2006. The goal was

Table 13 Summary of cadmium non-compliances in reports from DGAL monitoring programs (2002–2005)

Scallops	Year of DGAL monitoring program	Fishing area	Cadmium test result (mg kg^{-1} fresh wt)	Cadmium confirmation result (mg kg^{-1} fresh wt) (AFSSA/LERQAP)
Chlamys	2005	Pertuis Breton	1.18	*1.26 ± 0.18*
varia	Total no. of	Pertuis Breton	1.65	*1.64 ± 0.23*
	scallop	Pertuis Breton	1.12	1.05 ± 0.21
	samples = 14	Pertuis Breton	1.12	1.07 ± 0.21
		Pertuis Breton	1.40	1.07 ± 0.21
		Quiberon Bay	1.54	*1.62 ± 0.23*
		Quiberon Bay	1.5	*1.56 ± 0.22*
		Quiberon Bay	1.06	1.13 ± 0.16
Aequipecten	2004	Western Channel	1.13	1.33
opercularis	Total no. of scallop samples = 3			
C. varia	2002	Not specified	1.5	1.7
	Total no. of scallop samples = 9		1.6	1.7

*Italic: Samples confirmed as non-compliant with Regulations (EC) 1881/2006 (EC 2006c), according to EC 2007

to check the level of contamination in problem areas and in the smaller neighboring area of Pertuis d'Antioche (French Atlantic coast). These samplings also resulted in four non-compliant results for cadmium, and they were confirmed by AFSSA/LERQAP (two from the Pertuis Breton (French Atlantic coast) and two from Arcachon Bay). Because of these results, imposition of possible management measures is under examination in collaboration between the French Directorate General for Health (DGS – *Direction Générale de la Santé*) and the Directorate for Marine Fisheries and Aquaculture (DPMA – *Direction des Pêches Maritimes et de l'Aquaculture*).

With respect to the specific DGAL monitoring programs conducted in 2009, the presence of lead (Pb), cadmium (Cd), and mercury (Hg) concentrations in the white and dark meats of 108 batches of crustaceans (lobsters, spider crabs, common crabs, swimming crabs and king crabs) was found. These organisms, under investigation by the French National Reference Laboratory (NRL) were collected in France, as well as marine gastropods (common winkles, common whelk, abalone, and murex), echinoderms (purple sea urchin and black sea cucumber) and tunicates (ascidians) (Noël et al. 2011, in press). The results show mean concentrations for crustacean white meat of 0.041, 0.132, and 0.128 mg kg^{-1} for Pb, Cd, and Hg, respectively. These values were always lower than the European legislation maximum level of 0.50 mg kg^{-1} Cd. The concentration in the dark meat of common crabs (mean concentration 11.8 mg kg^{-1} and maximum 14.3 mg kg^{-1}) is well above the observed levels for white meat. The results for gastropods, echinoderms, and tunicates show that the highest levels of Hg and Cd were found in murex, 0.185 mg kg^{-1} and 0.853 mg kg^{-1}, respectively, whereas the highest level of Pb was detected in ascidians (0.505 mg kg^{-1}). Hg and Pb concentrations were systematically below the maximum regulatory levels (0.5 mg Hg kg^{-1} and 1.5 mg Pb kg^{-1} wet wt). For Cd, only two samples of murex (2.09 ± 0.42 and 2.33 ± 0.46 mg kg^{-1}) exceeded the French maximum level of 2.0 mg kg^{-1} wet wt.

Other data on contaminants contained in marketed shellfish are presented in Tables 3 and 6.

4.2.4 European Data on Chemical Contamination of Shellfish

There is no specific EU reference laboratory (EU-RL) for monitoring chemical contaminants in shellfish. However, there are four EU-RLs that test for lead, cadmium, mercury, PAHs, dioxins, and PCBs in animal tissues as indicated in Regulation (EC) No. 776/2006 (EC 2006a).

Chemical contamination levels in shellfish are monitored in many national and international surveys. However, many of these are not published. It would be useful for many researchers and governmental personnel to bring these scientific data together in a single national or European database that could be accessed through the internet.

5 Impact on Humans

Health risks associated with chemical contaminants are difficult to assess, owing to the fact that many produce only long-term action (chronic risk), and such contaminants reach humans through so many different sources (food, water, air, occupational, etc.). To assess the health impact of contaminated shellfish consumed by humans, exposure has to be estimated from contamination levels and consumption data.

5.1 Consumption Data for the General Population (INCA 2 2009)

Data on food consumption for the general population (including consumers and non-consumers of shellfish) may be taken from the INCA 2 (2009) survey (*Enquête Individuelle et Nationale sur la Consommation Alimentaire*) conducted in 2005–2007 by the Food Consumption and Nutritional Epidemiology Unit (OCA-EN) at AFSSA. In this survey, respondents recorded all types of food intake over a period of one full week. To account for seasonal effects, the survey was carried out in four phases spread over a period of 1 year. Food consumption data were obtained from consumption diaries that respondents kept over the targeted seven consecutive day period; in their diaries, respondents identified the foodstuffs and portions that were shown in a booklet of photographs (Suvimax 2002). The survey included 4000 adults and children that were representative of the French population. To ensure that the sample was nationally representative, it was stratified by region of residence and town size, and a quota method was used for age, sex, occupation, socio-occupational category, and size of household. The adult sample included 2624 individuals aged 18 and over. A special method was used to exclude bias resulting from under-estimation of food consumption by some respondents; those for whom the ratio between calories consumed and the basal metabolism, calculated using the Schofield method, was below a certain threshold were excluded from the calculations (706 were excluded). The collection of "normal" adults thus included 1918 individuals. The sample of children included 1455 individuals aged 3–17. This sample was not adjusted, because there was no formula for identifying low food-consuming sub-jects among children. In this survey, only the edible parts of foodstuffs were used to establish quantities consumed. The food groups counted as "solid foods" included all food groups in the INCA 2 (2009) nomenclature except for milk, water, soft drinks, alcoholic drinks, hot drinks, and soups.

At the most detailed level of the INCA 2 (2009) nomenclature, the reliability of the data for foods such as mollusks is not certain, because consumption was recorded for only 1 week. Amounts consumed were very low (Table 14). For com-parison, Table 14 gives data on the percent consumption of meat and fish by INCA 2 (2009) survey respondents. Mean daily consumption of shellfish, in the general population, was estimated to be 4.5 g in adults; this value varied widely by region and season of the year. Using this consumption level, shellfish consumption repre-sented 0.16% of overall solid food intake. However, the INCA 2 survey (2009) was not well suited to estimating shellfish consumption, because it included only a small number of shellfish consumers.

In conclusion, consumption of bivalve mollusks in France contributes little to the general population's overall food intake. Notwithstanding this conclusion, for the sake of regular shellfish consumers, continuing vigilance is necessary.

Table 14 Daily human consumption (grams per day of product consumed) according to the 2007 INCA 2 survey (INCA 2 2009)

	Adults (normal estimators) ($N = 1918$; aged 18 and older)				Children ($N = 1455$; aged 3–17)			
	Percentage of con-sumers	Mean	Standard deviation	Median	Percentage of con-sumers	Mean	Standard deviation	Median
Meat	92.0	49.7	37.5	42.4	91.5	38.1	28.8	32.9
Fish	79.3	26.5	24.7	21.2	78.7	18.3	17.6	14.3
Mollusks and crus-taceans	33.5	4.5	9.3	0	17.9	1.4	5.1	0

5.2 Consumption Data for High Consumers of Seafood (CALIPSO)

After the first INCA study (INCA1 1999), a specific work, called CALIPSO, was devoted to high consumers, i.e., adults who eat fish or seafood products at least twice a week (Leblanc et al. 2006). In Table 15 we present the data on mollusk consumption among high consumers of seafood products that were included in the CALIPSO survey ($n = 1011$ adults, including 246 men aged 18–64, 641 women aged 18–64, and 124 persons aged 65 and over). The results are given as means across the four sites studied, without distinction for age or gender (Leblanc et al. 2006).

Consumed bivalve species included cockle, mussel, king scallop, queen scallop, other scallops, razor clam, *Ruditapes* clams, other clams, oysters, warty venus, and wedge shell. Consumed gastropods included winkles, whelks, abalones, and limpets. The only echinoderm eaten in France is the sea urchin and the only tunicate eaten is the sea squirt.

In CALIPSO, mean consumption of bivalve mollusks among adults is estimated at 153 g week^{-1} (8 kg yr^{-1}). The highest mean consumption is for king scallops (39 g week^{-1}), followed by oysters (34 g week^{-1}) and mussels (22 g week^{-1}).

Table 15 Detail of mollusk consumption by "high consumers" of seafood in the CALIPSO survey (Leblanc et al. 2006). Data given in grams per week of fresh flesh

Mollusks	Mean (g week^{-1})	P5	P50	P95
Bivalves	119.7	7.5	79.8	350.3
Clam	0.2	0	0	0
Cockle	3.1	0	0	15.0
King scallop	39.3	0	25.0	156.3
Razor clam	0.4	0	0	0
Oyster	34.4	0	18.0	144.0
Mussel	22.5	0	17.5	70.0
Palourde clam	2.8	0	0	12.3
Other scallops	14.7	0	0	56.3
Warty venus	1.5	0	0	7.5
Wedge-shell, olive	0.3	0	0	0
Queen scallop	0.5	0	0	0
Gastropods	21.2	0	3.8	87.5
Winkle	4.2	0	0	25.0
Whelk	15.4	0	0	75.0
Abalone	0.6	0	0	0
Limpet	1.0	0	0	0
Echinoderms	11.6	0	0	52.5
Sea urchin	11.6	0	0	52.5
Tunicates	1.0	0	0	0
Sea squirt	1.0	0	0	0
All	153.5	10.0	106.1	413.5

Overall, these high consumers of seafood products eat, on average, twice the quantity of bivalve mollusks as do the shellfish consumers in the general population (INCA 2 2009); this is about the same level as the mean consumption of fish in the general population.

5.3 Exposure to Contaminants via Shellfish Consumption

5.3.1 Cadmium

In France, CALIPSO data show that the mean cadmium intake from shellfish is 1.26 µg week^{-1} in adults (Leblanc et al. 2005, 2006), which is about half of the new tolerable weekly intake (TWI) value that was recently revised by the European Food Safety Authority (EFSA); this value was revised from 7 to 2.5 µg kg^{-1} bwt week^{-1} (EFSA 2009). A recent PTMI (provisional tolerable monthly intake) value was given by JECFA (2010b) (25 µg kg^{-1} body wt month^{-1}); this value corresponds closely to a PTWI value of 5.3 µg kg^{-1} bwt week^{-1}. Shellfish consumed by adult men, who are high seafood consumers, lead to a cadmium intake of more than twice that of the average total intake from food in non-smoking adult men (EAT total diet survey by AFSSA). The cadmium intake varies from 8.2, 10, and 23% of the PTWI, depending on the threshold that is selected (P1, or P2 or P3 as defined in Table 3). The contribution of shellfish differs between regions. The shellfish that contribute most to cadmium intake by humans (CALIPSO survey) are king scallops (14% in Le Havre and 20% in Toulon), whelks (21%), scallops (19%), and oysters (11% in La Rochelle) (Leblanc et al. 2006).

5.3.2 Lead

In France, the mean intake of lead has fallen considerably in recent years. In 2005, the EAT survey indicated an average intake from food of 18 µg day^{-1} per adult, which amounts to 7% of the PTWI value set by the Joint FAO/WHO Expert Committee on Food Additives (JECFA) in 1986 (Leblanc et al. 2005). The mean lead intake from shellfish (CALIPSO survey) is 0.26 µg kg bwt^{-1} day^{-1} in adults (from Leblanc et al. 2005, 2006). However, in June 2010, the JECFA concluded that the PTWI could no longer be considered health protective and withdrew it (JECFA 2010b). EFSA came to the same conclusion in its opinion of March 2010 (EFSA 2010). Consumption of seafood (fresh fish, crustaceans, and mollusks) accounts for 3–11% of lead intake from total food. Shellfish contribute 0.7 µg day^{-1} of that intake. According to the CALIPSO survey, the main shellfish concerned are king scallops in Le Havre (22%), mussels in La Rochelle (16%), and sea urchins in Toulon (14%) (Leblanc et al. 2006).

5.3.3 Mercury

In seafood products, mercury is mainly present as methylmercury (see Section 5.4). For methylmercury (MeHg), both EFSA and AFSSA have acknowledged that some population groups are particularly at risk: pregnant and breast-feeding women, very young children, and fishing communities in heavily contaminated areas (EFSA 2004a; AFSSA 2004). Both agencies recommend that special information be aimed at these groups to encourage them to eat a wider range of fish species. In France, exposure study results show that values are two times lower than the PTWI (of 4 μg) for inorganic Hg kg^{-1} body weight. In adult males, who are high seafood consumers, the CALIPSO data suggest that shellfish result in an average intake of 0.47 μg day^{-1} of MeHg per adult, which approaches 1.2% of the PTWI (Leblanc et al. 2006). In general, fish contribute 86%, and mollusks and crustaceans 13% of MeHg exposure (EAT 2004; Leblanc et al. 2005, 2006; Sirot et al. 2008).

5.3.4 Arsenic

In 2003, the mean total arsenic intake in Europe was estimated at 125 μg day^{-1} in adults (SCOOP 2004); seafood accounted for over 50% of this exposure. The mean arsenic intake from shellfish from CALIPSO data is 84 μg kg bwt^{-1} $week^{-1}$ in adults (Leblanc et al. 2005, 2006). The seafood that contribute most to the French population's inorganic arsenic exposure are king scallops (8.6% of intake from seafood) and oysters (7.0%) (Sirot et al. 2009). In the general population, shellfish contribute 0.2% of the PTWI for total arsenic (EAT 2004; Leblanc et al. 2005). However, it was noted in the 72nd JECFA committee meeting that the PTWI of 15 μg kg bwt^{-1} (equivalent to 2.1 μg kg bwt^{-1} day^{-1}) approaches the benchmark dose lower limit ($BMDL_{05}$), and therefore the PTWI is no longer appropriate. The committee withdrew the previous PTWI (JECFA 2010a). EFSA concluded that the overall range of $BMDL_{01}$ values of 0.3–8 μg kg^{-1} bwt day^{-1} should be used, instead of a single reference point, in characterizing the risk of inorganic arsenic (EFSA 2009).

5.3.5 Organostannic Compounds

In France, the average exposure of high seafood consumers to nine organostannic compounds is far below the tolerable daily intake. This intake is 8–19% of the TDI of 0.1 μg Sn kg^{-1} bwt set by EFSA for the combined total from the following Sn compounds: tributyltin (TBT), dibutyltin (DBT), triphenyltin (TPT), and dioctyltin (DOT) (AFSSA 2006; EFSA 2004b; Guérin et al. 2007).

5.3.6 Dioxins

The mean dioxin intake from shellfish for the sum of PCDD/Fs and DL-PCBs is 18.7 $μg.kg^{-1}$ bwt $week^{-1}$ in adults (from CALIPSO data; Leblanc et al. 2006). However, it is important to note that these values are overestimated, because cooking seafood reduces the PCDD content (Hori et al. 2005). The shellfish contribution to the tolerable intake is low (5.73% for all species).

5.3.7 PCBs

Only 28% of high seafood consumers show indicator PCB (sum of PCB 28, 52, 101, 118, 153, and 180) levels below the TDI of 0.02 μg kg–1 bwt day–1, the average being 0.40 $\pm$ 0.55 μg kg–1 bwt day–1. Shellfish contribute only 9.5% to the TDI, 45% of which comes from the king scallop (Leblanc et al. 2006).

5.3.8 Body Burdens for These Trace Elements

The CALIPSO survey provides data on the body burdens (saturation) of high seafood consumers (adults) for lead, cadmium, mercury and arsenic. However, from these data, it is not possible to determine how much shellfish contribute to the body burden. Concentrations of these chemical contaminants, measured in blood and urine, are compared with a "basal value." This basal value is defined as the value found at the 95th percentile (P95) of the general French population that is not occupationally exposed. This value should not be interpreted as a maximum allowable quantity, but it makes it possible to identify a possible body overload. In conclusion, high seafood consumers do not display a significantly higher body burden than does the P95 of the general population for lead, cadmium, or mercury. For lead, 6% of high seafood consumers exceed the basal value of 90 μg L^{-1} for men and 70 μg L^{-1} for women. There were no observed blood concentration of lead >200 μg L^{-1}, the concentration above which a person is put under medical observation. For cadmium, fewer than 5% of individuals retained cadmium levels in urine higher than the basal value of 2 mg kg^{-1} creatinine. For mercury, only 3% of the values exceeded the basal value of 10 μg L^{-1} in blood. No signs of health risk impairment were identified for any of these three contaminants. However, 22% of individuals displayed inorganic arsenic levels that exceeded the basal concentration in urine of 10 mg kg^{-1} creatinine, which is the P95 value for the general population (INRS 2010; Pillière and Conso 2007).

We conclude from the foregoing that, for high seafood consumers, the contribution of shellfish to inorganic contaminants was 1–10% of TWI or PTWI for Cd, MeHg, and Sn (up to 19% for Sn), and the arsenic body burden was higher for 22% of individuals studied. These percentages will differ if the established effective regulatory threshold is different (Table 3).

5.4 Health Risk Assessment Uncertainty from Contaminant Bioavailability and Speciation Effects

The regulatory limits for lead, cadmium, and mercury that were established in Regulation (EC) 1881/2006 (EC 2006c) are based on the total concentrations of them that exist in foodstuffs. However, only the bioavailable fraction can be transferred from shellfish to humans, during digestion. This fraction is influenced by several factors and is rarely 100% of the amount present.

For mercury, it is the methylated form that predominates as a seafood residue, and this organic form is also the most toxic (Nakagawa et al. 1997; Storelli et al. 1998). A study of oysters and mussels sampled in 1996 under the RNO sampling program shows MeHg/THg (total Hg) ratios ranging from 11 to 88%. No notable differences were observed between the two mercury species, but there was considerable geographical variability (Claisse et al. 2001). Bioamplification has been observed in organic forms of mercury, with an increase in concentration at each trophic step in the food chain. In the CALIPSO survey, MeHg/THg ratios in shellfish ranged from 50 to 100% (Leblanc et al. 2006).

The toxicity of arsenic depends on its chemical form and its bioavailability. Inorganic forms of arsenic are more toxic than are the organic forms (Michel 1993; Sharma and Sohn 2009). A high proportion of the organic arsenic in seafood is in weakly toxic forms such as arsenobetaine and trimethylarsine. These forms are rapidly excreted (ATSDR 2007; Liber et al. 2006). According to the WHO, there are some (but limited) data showing that 25% of total arsenic in foodstuffs is in inorganic form. The data from a French study (Noël et al. 2003) suggest that, in fishery products, 5–10% of arsenic is in inorganic form, whereas the CALIPSO study gives figures ranging from 0.1 to 3.5% in fish and from 0.1 to 6.7% in shellfish (Sirot et al. 2009). However, the percentage of inorganic arsenic is quite variable in fish and shellfish, and data from the international literature indicate that the percentage of inorganic arsenic in marine/estuarine finfish does not exceed 7.3%. However, in shellfish, it can reach 25% in organisms from presumably uncontaminated areas, although there are few data available for freshwater organisms. However, percentages can be much higher in organisms from contaminated areas and in seaweed (Schoof and Yager 2007; Lorenzana et al. 2009).

To determine the public health risk due to mercury or arsenic, it is thus necessary to know the inorganic/organic proportions and not only the total levels.

Efforts have been made in various studies to quantify the bioavailability, or rather bioaccessibility, of trace elements that are accumulated by bivalves (He et al. 2010; Metian et al. 2009). Amiard et al. (2008) simulated human digestion in vitro with the flesh of naturally contaminated oysters, whelks, mussels, scallop species, and *Ruditapes* clams. The total concentrations in these samples exceeded regulatory limits for the following (Amiard et al. 2008):

- Cd in whelks (*B. undatum*) purchased in France and in the adductor muscles of noble scallops (*Chlamys nobilis*) from Hong Kong;
- Pb in oysters (*O. edulis*) from Restronguet Creek, UK, and Zn in whelks, and
- Cu and Zn in all samples of oysters from contaminated sites.

However, these comparisons are based on Australian and Asian standards that Europe does not recognize. If the concentrations recorded were indeed bioaccessible concentrations, only the levels of Cd in scallop species and Zn in whelks would be acceptable.

Although levels of arsenic in the urine and more specifically inorganic arsenic are satisfactory biomarkers for occupational and drinking water exposures, the literature data show that consumption of seafood gives variable results. The amount of total or inorganic arsenic in the urine is, therefore, not a relevant or usable indicator of the intensity of exposure to the most toxic forms of arsenic ingested with food, and with seafood in particular. To assess the health risk of ingesting arsenic via seafood, the species of arsenic must be taken into account, because there are significant differences in toxicity among the different chemical species. For example, the mean LD_{50} (lethal dose, 50%) in rats, expressed in mg kg^{-1} bwt, is 14 for potassium arsenite, 20 for calcium arsenate, 700–1800 for MMA (monomethylarsonic acid), 700–2600 for DMA (dimethylarsinic acid), and over 10,000 for arsenobetaine. In drinking water, arsenic is mainly found in the inorganic form as the oxide anions arsenite and arsenate. The main foodstuffs supplying inorganic arsenic are cereals, flour, and raw rice (Schoof et al. 1999), but seafood contain several organic arsenic compounds and are a major food source of arsenic (Francesconi and Edmonds 1998; Munoz et al. 2000). Arsenic in fish, most shellfish, and many crustaceans is mainly in the form of arsenobetaine, whose very weak toxicity has been established (Kaise et al. 1985; Sabbioni et al. 1991). Arsenobetaine is quickly excreted in an unaltered form in the urine (70% in 3 days) (Cannon et al. 1983) and does not react with the reagents used in urinary tests. Hence, arsenobetaine is clearly differentiated during arsenic speciation in the urine, and several experimental studies have shown that its consumption does not significantly alter the parameters of urine analyses for inorganic arsenic (Buchet et al. 1996; Heinrich-Ramm et al. 2002; Hsueh et al. 2002).

Algae, bivalves, crustaceans, and fish all contain derivatives of ribose and arsenic called arsenoribosides (arsenosugars), which are metabolized and excreted in the urine, particularly as DMA(V) and in the form of dimethyloxarsylethanol and trimethylarsine oxide (Francesconi et al. 2002; Ma and Le 1998; Wei et al. 2003). It has been observed that ingestion of arsenoribosides via food invalidates the use of urine testing for inorganic arsenic derivatives, and as an exposure marker for these derivatives. As a result, these tests cannot satisfactorily reflect intake of inorganic arsenic (for which there is a risk of excess cancers) in individuals consuming seafood (Borak and Hosgood 2007; Heinrich-Ramm et al. 2002; Ma and Le 1998). Considering that organic arsenic from seafood is usually eliminated within 3 days (Crecelius 1977; Freeman et al. 1979), it is recommended that urine tests for inorganic arsenic should be performed at least 3 or 4 days after any seafood consumption (Foa et al. 1984; Kales et al. 2006).

As previously mentioned, health risks associated with chemical contaminants are difficult to assess because the risks they pose are normally of a chronic nature, and their sources of human exposure are numerous. Therefore, it is not possible to attribute a high body burden specifically to shellfish consumption, even though seafood is a major contributor of some contaminants, especially arsenic and mercury.

6 Conclusions

The major conclusions we have reached from compiling and reviewing the literature cogent to the topic of this chapter are as follows:

- Both organic and inorganic chemicals have been identified as residual chemical contaminants in shellfish. Some contaminants, particularly metals, dioxins, DL-PCBs, and PAH that appear as residues in mollusks, pose a potential hazard to consumers, which has resulted in European regulatory limits being established for them.
- To protect shellfish from chemical contamination, shellfish production and commercialization are managed according to safe practices stipulated by the European "Hygiene Package" regulations. Product quality is maintained by controlling facilities, tracking major steps in shellfish production, and ensuring that defective batches are kept from the market. Such regulation ensures greater transparency and product quality for consumers. However, limits to regulating shellfish production also exist, because in France, it is difficult to trace all production steps of living shellfish from the earliest to latest stages, particularly for oysters, wherein the same oyster may be successively raised at facilities in different areas.
- Although monitoring results show few non-conformities, the samplings that are made cannot be considered as representing all shellfish production in France, because the number of samples taken is limited. In addition, when residue levels are exceeded, they normally occur in oysters and mussels, which are the most commonly eaten species. Hence, self-inspection by producers, enforcement of compliance with good practices, and regular checks on production are indispensable additional measures to ensure food safety.
- Last, but not least, is that the chemical monitoring network that has been set up in France (the RNO program and its successor ROCCH) to screen for contaminants clearly shows that there is low chemical contamination of mollusks and of seawater in which the mollusks live. Moreover, when shellfish contamination occurs, it poses a generally low risk to the general French population, because the proportion of the diet that shellfish constitutes is low. The exceptions are when contaminants reach those people who are either high consumers of shellfish or a more susceptible population, such as pregnant and breast-feeding women and very young children. Appropriate research programs should first be developed to protect these more susceptible categories of the population.

To improve the safety of consuming shellfish potentially contaminated with chemicals, the following suggestions are given:

- The relaying (depuration) program currently used to purify shellfish of microbiological contamination before commercialization should be further researched to determine if and how chemical residues in shellfish could be similarly reduced before consumption. The alternative that has been used to date, i.e., closing contaminated areas for long time periods, results in significant economic losses.

The monitoring of farmed shellfish should be extended to other chemicals that are suspected to present a consumer risk (especially arsenic, which the CALIPSO study disclosed to have high consumer urine levels, and cadmium, which was detected at abnormal levels in some shellfish in 2009). We would also suggest monitoring for TBT and PAH contamination levels to ensure that these chemicals do not migrate from the harbor to oyster farms, as was observed to occur in Arcachon Bay. Other monitoring candidates in estuaries would include cadmium and PCBs, which can pose serious problems for the sale of some shellfish.

Finally, from the data assembled in this review, we conclude that there is a strong argument not to curtail existing monitoring programs in edible shellfish. The major reason for continuing monitoring activities is that great variability exists in the magnitude to which different contaminants in shellfish bioconcentrate. Both environmental and species parameters are known to affect the degree of bioconcentration and bioaccumulation of potentially harmful residues, and, moreover residue loads are affected by the season during which the shellfish are harvested. Therefore, under equal conditions of environmental contamination, some species do exceed the European regulatory limits, whereas others do not. These variabilities explain the necessity of why monitoring was extended to farmed shellfish species by the ROCCH and why monitoring activities should continue.

7 Summary

In this review, we address the identification of residual chemical hazards in shellfish collected from the marine environment or in marketed shellfish. Data, assembled on the concentration of contaminants detected, were compared with the appropriate regulatory and food safety standards. Moreover, data on human exposure and body burden levels were evaluated in the context of potential health risks.

Shellfish farming is a common industry along European coasts. The primary types of shellfish consumed in France are oysters, mussels, king scallops, winkles, whelks, cockles, clams, and other scallops. Shellfish filter large volumes of water to extract their food and are excellent bioaccumulators. Metals and other pollutants that exist in the marine environment partition into particular organs, according to their individual chemical characteristics. In shellfish, accumulation often occurs in the digestive gland, which plays a role in assimilation, excretion, and detoxification of contaminants. The concentrations of chemical contaminants in bivalve mollusks are known to fluctuate with the seasons.

European regulations limit the amount and type of contaminants that can appear in foodstuffs. Current European standards regulate the levels of microbiological agents, phycotoxins, and some chemical contaminants in food. Since 2006, these regulations have been compiled into the "Hygiene Package." Bivalve mollusks must comply with maximum levels of certain contaminants as follows: lead (1.5 mg kg^{-1}), cadmium (1 mg kg^{-1}), mercury (0.5 mg kg^{-1}), dioxins (4 pg g^{-1} and dioxins + DL-PCBs 8 pg g^{-1}), and benzo[a]pyrene (10 μg kg^{-1}).

In this review, we identify the levels of major contaminants that exist in shellfish (collected from the marine environment and/or in marketed shellfish). The following contaminants are among those that are profiled: Cd, Pb, Hg, As, Ni, Cr, V, Mn, Cu, Zn, Co, Se, Mg, Mo, radionuclides, benzo[a]pyrene, PCBs, dioxins and furans, PAHs, TBT, HCB, dieldrin, DDT, lindane, triazines, PBDE, and chlorinated paraffins.

In France, the results of contaminant monitoring have indicated that Cd, but not lead (< 0.26 mg kg^{-1}) or mercury (< 0.003 mg kg^{-1}), has had some non-compliances. Detections for PCBs and dioxins in shellfish were far below the regulatory thresholds in oysters (< 0.6 pg g^{-1}), mussels (< 0.6 pg g^{-1}), and king scallops (< 0.4 pg g^{-1}). The benzo[a]pyrene concentration in marketed mussels and farmed shellfish does not exceed the regulatory threshold. Some monitoring data are available on shellfish flesh contamination for unregulated organic contaminants.

Of about 100 existing organostannic compounds, residues of the mono-, di-, and tri-butyltin (MBT, DBT, and TBT) and mono-, di-, and triphenyltin (MPT, DPT, and TPT) compounds are the most frequently detected in fishery products. Octyltins are not found in fishery products. Some bivalve mollusks show arsenic levels up to 15.8 mg kg^{-1}. It seems that the levels of arsenic in the environment derive less from bioaccumulation, than from whether the arsenic is in an organic or an inorganic form. In regard to the other metals, levels of zinc and magnesium are higher in oysters than in mussels.

To protect shellfish from chemical contamination, programs have been established to monitor water masses along coastal areas. The French monitoring network (ROCCH) focuses on environmental matrices that accumulate contaminants. These include both biota and sediment. Example contaminants were studied in a French coastal lagoon (Arcachon Bay) and in an estuary (Bay of Seine), and these were used to illustrate the usefulness of the monitoring programs. Twenty-one pesticidal and biocidal active substances were detected in the waters of Arcachon Bay during the summers from 1999 to 2003, at concentrations ranging from a few nanograms per liter to several hundred nanograms per liter. Most of the detected substances were herbicides, including some that are now banned. Organotin compounds have been detected in similarly semi-enclosed waters elsewhere (bays, estuaries, and harbors). However, the mean concentrations of cadmium, mercury, lead, and benzo[a]pyrene, in transplanted mussels, were below the regulatory limits.

In 2007, the mean daily consumption of shellfish in the general French population was estimated to be 4.5 g in adults; however, a wide variation occurs by region and season (INCA 2 study). Tabulated as a proportion of the diet, shellfish consumption represents only 0.16% of overall solid food intake. However, the INCA 2 survey was not well suited to estimating shellfish consumption because of the small number of shellfish consumers sampled. In contrast, the mean consumption rate of bivalve mollusks among adult high consumers of fish and seafood products, i.e., adults who eat fish or seafood at least twice a week, was estimated to be 153 g week^{-1}

(8 kg yr^{-1}). The highest mean consumption is for king scallops (39 g week^{-1}), followed by oysters (34 g week^{-1}) and mussels (22 g week^{-1}). Thus, for high seafood consumers, the contribution of shellfish to inorganic contaminant levels is 1–10% of TWI or PTWI for Cd, MeHg, and Sn (up to 19% for Sn), and the arsenic body burden is higher for 22% of individuals studied.

The human health risks associated with consuming chemical contaminants in shellfish are difficult to assess for several reasons: effects may only surface after long-term exposure (chronic risk), exposures may be discontinuous, and contamination may derive from multiple sources (food, air, occupational exposure, etc.). Therefore, it is not possible to attribute a high body burden specifically to shellfish consumption even if seafood is a major dietary contributor of any contaminant, e.g., arsenic and mercury.

The data assembled in this review provide the arguments for maintaining the chemical contaminant monitoring programs for shellfish. Moreover, the results presented herein suggest that monitoring programs should be extended to other chemicals that are suspected of presenting a risk to consumers, as illustrated by the high concentration reported for arsenic (in urine) of high consumers of seafood products from the CALIPSO study. In addition, the research conducted in shellfish-farming areas of Arcachon Bay highlights the need to monitor TBT and PAH contamination levels to ensure that these chemical pollutants do not migrate from the harbor to oyster farms.

Finally, we have concluded that shellfish contamination from seawater offers a rather low risk to the general French population, because shellfish do not constitute a major contributor to dietary exposure of chemical contaminants. Notwithstanding, consumer vigilance is necessary among regular shellfish consumers, and especially for those residing in fishing communities, for pregnant and breast-feeding women, and for very young children.

References

Afssa (2004) Avis du 16 mars 2004 de l'Agence relatif à la réévaluation des risques sanitaires du méthylmercure liés à la consommation des produits de la pêche au regard de la nouvelle dose hebdomadaire tolérable provisoire (DHTP). Available from http://www.afssa.fr/Documents/RCCP2003sa0380.pdf

Afssa (2005) Rapport sur "Dioxines, furanes et PCB de type dioxines: evaluation de l'exposition de la population française", Novembre 2005, 57 pages. Available from http://www.afssa.fr/Documents/RCCP2005sa0372.pdf

Afssa (2006) Avis du 18 avril 2006 sur l'évaluation des risques liés à la présence d'organoétains dans les aliments. Available from http://www.afssa.fr/Documents/RCCP2005sa0091.pdf

Afssa (2007a) Avis du 26 juillet 2007 sur la mise en place de règles hygiéniques d'utilisation de l'eau de mer propre pour la manipulation des produits de la mer. Available from http://www.afssa.fr/Documents/RCCP2006sa0314.pdf

Afssa (2007b) Avis du 31 octobre 2007 sur la pertinence d'établir une teneur maximale en cadmium pour les gastéropodes, les échinodermes et les tuniciers et à l'évaluation des risques sanitaires liés à des teneurs élevées en cadmium dans les bulots et les pétoncles. Available from http://www.afssa.fr/Documents/RCCP2007sa0098.pdf

Afssa (2008a) Avis du 21 Mars 2008 relatif à l'évaluation du dispositif de surveillance du milieu et à l'évaluation du risque lié à la consommation des coquillages, notamment dans la situation du bassin d'Arcachon. Available from http://www.afssa.fr/Documents/RCCP2006sa0254b.pdf

Afssa (2008b) Rapport sur l'Evaluation de la surveillance chimique des zones de production conchylicole du risque lié à la consommation des coquillages, notamment dans la situation du bassin d'Arcachon. Available from http://www.afssa.fr/Documents/RCCP2006sa0254bRa. pdf

Afssa (2008c) Avis du 16 septembre 2008 sur la pertinence des outils de détection des phycotoxines lipophiles dans les coquillages. Available from http://www.anses.fr/cgi-bin/countdocs.cgi?Documents/RCCP2008sa0268.pdf

Alzieu C, Michel P, Tolosa I, Bacci E, Mee LD, Readman JW (1991) Organotin compounds in the Mediterranean: a continuing cause for concern. Mar Environ Res 32:261–270

Amiard JC (1991) Réponses des organismes marins aux pollutions métalliques. In: CNRS (ed) Réactions des êtres vivants aux changements de l'environnement. Actes des Journées de l'Environnement du CNRS, Paris, pp 197–205

Amiard JC, Amiard-Triquet C, Barka S, Pellerin J, Rainbow PS (2006) Metallothioneins in aquatic invertebrates: their role in metal detoxification and their use as biomarkers. Aquat Toxicol 76: 160–202

Amiard JC, Amiard-Triquet C, Berthet B, Metayer C (1986) Contribution to the ecotoxicological study of Cd, Pb, Cu and Zn in the mussel *Mytilus edulis*. 1 – Field study. Mar Biol 90:425–431

Amiard JC, Amiard-Triquet C, Berthet B, Metayer C (1987) Comparative study of the patterns of bioaccumulation of essential (Cu, Zn) and non-essential (Cd, Pb) trace metals in various estuarine and costal organisms. J Exp Mar Biol Ecol 106:73–89

Amiard JC, Amiard-Triquet C, Charbonnier L, Mesnil A, Rainbow PS, Wang WX (2008) Bioaccessibility of essential and no-essential metals in commercial shellfish from Western Europe and Asia. Food Chem Toxicol 46:2010–2022

Amiard JC, Berthet B (1996) Fluctuations of cadmium, copper, lead and zinc concentrations in field populations of the Pacific oyster *Crassostrea gigas* in the Bay of Bourgneuf (Atlantic coast, France). Ann Inst Oceanogr 72:195–207

Amiard-Triquet C, Amiard JC (1980). Radioécologe des milieux aquatiques. Masson, Paris, 191 p

Andral B, Stanisiere JY, Sauzade D, Damier E, Thebault H, Galgani F, Boissery P (2004) Monitoring chemical contamination levels in the Mediterranean based on the use of mussel caging. Mar Pollut Bull 49:704–712

Apeti DA, Lauenstein GG, Christensen JD, Kimbrough K, Johnson WE, Kennedy M, Grant KG (2010) A historical assessment of coastal contamination in Birch Harbor, Maine based on the analysis of mussels collected in the 1940s and the Mussel Watch Program. Mar Pollut Bull 60:732–742

ATSDR (2007) ToxGuide for Arsenic. cas#7440-38-2. http://www.atsdr.cdc.gov/toxguides/toxguide-2.pdf?id=21%26;tid=3

Auby I, Maurer D (2004) Etude de la reproduction des huîtres creuses dans le bassin d'Arcachon. Ifremer Edition, Nantes, 327 p

Averty B, Michel P, Chiffoleau JF (2005) Les composés organostanniques dans les mollusques du littoral français. RNO, 2005. Ifremer Editor, Nantes, pp 35–38

Bard SM (2000) Multixenobiotic resistance as a cellular defense mechanism in aquatic organisms. Aquat Toxicol 48:357–389

Berthet B (2008) Les espèces sentinelles. In: Amiard J-C, Amiard-Triquet C (eds) Les biomarqueurs dans l'évaluation de l'état écologique des milieux aquatiques. Lavoisier, Tec&Doc, Paris, pp 121–148

Borak J, Hosgood D (2007) Seafood arsenic: implications for human risk assessment. Regul Toxicol Pharmacol 47:204–212

Bryan GW (1976) Heavy metal contamination in the sea. In: Johnston R (ed) Marine pollution. Academic, London, pp 185–302

Bryan GW, Langston WJ, Humerstone LG, Burt GR, Ho YB (1983) An assessment of the gastropod *Littorina littorea* as an indicator of heavy-metal contamination in United kingdom estuaries. J Mar Biol Ass 63:327–345

Buchet JP, Lison D, Ruggeri M, Foa V, Elia G (1996) Assessment of exposure to inorganic arsenic, a human carcinogen, due to the consumption of seafood. Arch Toxicol 70:773–778

Bügel SH, Sandström B, Larsen EH (2001) Absorption and retention of selenium from shrimps in man. J Trace Elem Med Biol 14:198–204

Butler PA (1973) Residues in fish, wildlife, and estuaries; organochlorine residues in estuarine molluscs, 1965–1972. National Pesticides Monitoring Program. J Pestic Monit 6:238–362

Cannon JR, Saunders JB, Toia RF (1983) Isolation and preliminary toxicological evaluation of arsenobetaine – the water soluble arsenical constituent from the hepatopancreas of the western rock lobster. Sci Total Environ 31:181–185

Cantillo AY (1998) Comparison of results of mussel watch programs of the United States and France with worldwide mussel watch studies. Mar Pollut Bull 36:712–717

Casas S, Bacher C (2006) Modelling trace metal (Hg and Pb) bioaccumulation in the Mediterranean mussel, *Mytilus galloprovincialis*, applied to environmental monitoring. J Sea Res 56:168–181

Chiffoleau J-F, Auger D, Chertier E, Le Goff R, Justome V, Maheux F, Pierre-Duplessis O, Etourneau C (2002) Variabilité de la contamination des Bulots et Coquilles Saint Jacques en Baie de Seine par les métaux. Rapport Scientifique du programme Seine-Aval, Phase 2

Chong K, Wang WX (2001) Comparative studies on the biokinetics of Cd, Cr, and Zn in the green mussel *Perna viridis* and the Manila clam *Ruditapes philippinarum*. Environ Pollut 115:107–121

Claisse D (1989) Chemical contamination of French coasts: the results of a ten years mussel watch. Mar Pollut Bull 20:523–528

Claisse D (1992) Accumulation des métaux lourds et polluants organiques par les coquillages. In: Lesne J (ed) Coquillages et santé publique. Du risque à la prévention. ENS; Rennes (Pacé), pp 99–111. Available from http://archimer.ifremer.fr/doc/00017/12785/

Claisse D (1999) Le RNO: programmes actuels. Surveillance du Milieu Marin. Travaux du réseau national d'observation de la qualité du milieu marin: Bulletin RNO Edition 1999, pp 5–10

Claisse D, Alzieu C (1993) Copper contamination as a result of antifouling paint regulations? Mar Pollut Bull 26:395–397

Claisse D, Cossa D, Bretaudeau-Sanjuan J, Touchard G, Bombled B (2001) Methylmercury in molluscs along the French coast. Mar Pollut Bull 42:329–332

Claisse D, Le Moigne M, Durand G, Beliaeff B (2006) Ligne de base: les contaminants chimiques dans les huîtres et les moules du littoral français. Surveillance du Milieu Marin. Travaux du réseau national d'observation de la qualité du milieu marin: Bulletin RNO Edition 2006, pp 27–51

CNC (2010) http://www.cnc-france.com/index.php?rub=2

Crecelius EA (1977). Changes in the chemical speciation of arsenic following ingestion by man. Environ Health Perspect 19:47–50

De Kock WC, Kramer KJM (1994) Active biomonitoring (ABM) by translocation of bivalve molluscs. In: Kramer KJM (ed) Biomonitoring of coastal waters and estuarines. CRC Press, Boca Raton, FL, pp 51–84

Devier MH, Augagneur S, Budzinski H, Le Menach K, Mora P, Narbonne JF, Garrigues P (2005) One-year monitoring survey of organic compounds (PAHs, PCBs, TBT), heavy metals and biomarkers in blue mussels from the Arcachon bay, France. J Environ Monit 7:224–240

EAT (2004) Etude de l'alimentation totale française. Mycotoxines, minéraux et éléments traces. Leblanc JC (coord.) Verger P, Guérin T, Volatier JL (INRA/DGAL, Mai 2004, 68 pages)

EC (2000) Directive (EC) No 60/2000 of the European parliament and of the council of 23 October 2000 establishing a framework for Community action in the field of water policy. Off J Eur Communities L327:1–72

EC (2001) Regulation (EC) N°466/2001 of the European parliament and the council of 8 March 2001 setting maximum levels for certain contaminants in foodstuffs. Off J Eur Communities L77:1–13

EC (2004a) Regulation (EC) No 853/2004 of the European parliament and the council of 29 April 2004 laying down specific hygiene rules for food of animal origin. Off J Eur Communities L226:22–82

EC (2004b) Regulation (EC) No 852/2004 of the European parliament and the council of 25 June 2004 on the hygiene of foodstuffs. Off J Eur Communities L226:3–21

EC (2004c) Regulation (EC) No 854/2004 of the European parliament and the council of 25 June 2004 laying down specific rules for the organisation of official controls on products of animal origin intended for human consumption. Off J Eur Communities L226:83–127

EC (2006a) Regulation (EC) No 776/2006 of the European parliament and the council of 23 May 2006 amending Annex VII to Regulation (EC) No 882/2004 of the European Parliament and of the Council as regards Community reference laboratories. Off J Eur Communities L136:3–8

EC (2006b) Regulation (EC) No 882/2006 of the European parliament and the council of 16 June 2006 concerning tenders notified in response to the invitation to tender for the export of common wheat issued in Regulation (EC) No 1059/2005. Off J Eur Communities L164:24–26

EC (2006c) Regulation (EC) No 1881/2006 of the European parliament and the council of 19 December 2006 setting maximum levels for certain contaminants in foodstuffs. Off J Eur Communities L364:5–22

EC (2007) Regulation (EC) No 333/2007 of 28 March 2007 laying down the methods of sampling and analysis for the official control of the levels of lead, cadmium, mercury, inorganic tin, 3-MCPD and benzo(a)pyrene in foodstuffs. Off J Eur Communities L88:29–38

EC (2008a) Regulation (EC) No 629/2008 of the European parliament and the council of 2 July 2008 amending Regulation (EC) No 1881/2006 setting maximum levels for certain contaminants in foodstuffs. Off J Eur Communities L173:6–9

EC (2008b) Directive (EC) No 105/2008 of the European parliament and of the council of 16 December 2008 on environmental quality standards in the field of water policy. Off J Eur Communities L348:84–97

EEC (1991) Directive (EEC) No 492/1991 of 15 July 1991 laying down the health conditions for the production and the placing on the market of live bivalve molluscs. 19 p. Available from: http://www.fao.org/fishery/legalframework/nalo_france/fr

EFSA (2004a) Opinion of the scientific panel on contaminants in the food chain on a request from the commission related to mercury and methylmercury in food. EFSA J 34:1–14

EFSA (2004b) Opinion of the scientific panel on contaminants in the food chain on a request from the commission to assess the health risks to consumers associated with exposure to organotins in foodstuffs. EFSA J 102:1–119

EFSA (2009) Scientific opinion on cadmium in food. EFSA panel on contaminants in the food chain (CONTAM). EFSA J 980:1–139

EFSA (2010) Scientific opinion on lead in food. EFSA J 8(4):1570

England GC, McGrath TP, Gilmer L, Seebold JG, Lev-On M, Hunt T (2001) Hazardous air pollutant emissions from gas-fired combustion sources: emissions and the effects of design and fuel type. Chemosphere 42:745–764

Ettajani H, Amiard-Triquet C, Jeantet AY, Amiard JC, Ballan-Dufrançais C (1996) Fate and effects of soluble or sediment-bound arsenic in oysters (*Crassostrea gigas* Thun.). Arch Environ Contam Toxicol 31:38–46

Fatta-Kassinos D, Meric S, Nikolaou A (2011) Pharmaceutical residues in environmental waters and wastewater: current state of knowledge and future research. Anal Bioanal Chem 399:251–275

Foa V, Colombi A, Maroni M, Buratti M, Calzaferri G (1984) The speciation of the chemical forms of arsenic in the biological monitoring of exposure to inorganic arsenic. Sci Total Environ 34:241–259

Francesconi KA, Edmonds JS (1998) Arsenic species in marine samples. Croat Chem Acta 71:343–359

Francesconi KA, Tangaar R, Mc Kenzie CJ, Goessler W (2002) Arsenic metabolites in human urine after ingestion of an arsenosugar. Clin Chem 48:92–101

Franco J, Borja Á, Solaun O, Pérez V (2002) Heavy metals in molluscs from the Basque Coast (Northern Spain): results from an 11-year monitoring programme. Mar Pollut Bull 44:973–976

Freeman HC, Uthe JF, Fleming RB, Odense PH, Ackman RG, Landry G, Musial C (1979) Clearance of arsenic ingested by man from arsenic contaminated fish. Bull Environ Contam Toxicol 22:224–229

Gagnaire B, Thomas-Guyon H, Burgeot T, Renault T (2006) Pollutant effects on Pacific oyster, *Crassostrea gigas* (Thunberg), hemocytes: screening of 23 molecules using flow cytometry. Cell Biol Toxicol 22:1–14

Geffard A, Amiard JC, Amiard-Triquet C (2002) Kinetics of metal elimination in oysters from a contaminated estuary. Comp Biochem Physiol C: Toxicol Pharmacol 131:281–293

George SG, Pirie, BJS, Cheyne AR, Coombs TL, Grant PT (1978) Detoxication of metals by marine bivalves: an ultrastructural study of the compartmentation of copper and zinc in the oyster, *Ostrea edulis*. Mar Biol 45:147–156

George SG, Pirie BJ, Frazier JM, Thomson JD (1984) Interspecies differences in heavy metal detoxication in oysters. Mar Environ Res 14:462–464

Goldberg ED (1975) The mussel watch – a first step in global marine monitoring. Mar Pollut Bull 6:111–113

Goldberg ED, Bowen VT, Farrington JW, Harvey G, Martin JH, Parker PL, Risebrough RW, Robertson W, Schneider E, Gamble E (1978) The mussel watch. Environ Conserv 5:101–125

Goldberg ED, Koide M, Hodge V, Flegal AR, Martin J (1983) U.S. mussel watch: 1977–1978 results on trace metals and radionuclides. Estuar Coastal Shelf Sci 16:69–93

Gomez-Ariza JL, Morales E, Giraldez I (1999) Uptake and elimination of tributyltin in clams, *Venerupis decussata*. Mar Environ Res 47:399–413

Guerin T, Sirot V, Volatier JL, Leblanc JC (2007) Organotin levels in seafood and its implications for health risk in high-seafood consumers. Sci Total Environ 388:66–77

Hamilton SJ (2004) Review of selenium toxicity in the aquatic food chain. Sci Total Environ 326: 1–31

Han BC, Jeng WL, Tsai YN, Jeng MS (1993) Depuration of copper and zinc by green oysters and blue mussels of Taiwan. Environ Pollut 82:93–97

He M, Ke CH, Wang WX (2010) Effects of cooking and subcellular distribution on the bioaccessibility of trace elements in two marine fish species. J Agric Food Chem 58:3517–3523

Hédouin L, Pringault O, Metian M, Bustamante P, Warnau M (2007) Nickel bioaccumulation in bivalves from the New Caledonia lagoon: seawater and food exposure. Chemosphere 66: 1449–1457

Heinrich-Ramm R, Mindt-Prüfert S, Szadkowski D (2002) Arsenic species excretion after controlled seafood consumption. J Chromatogr B 778:263–273

Heinzow B, Mohr S, Ostendorp G, Kerst M, Körner W (2007) PCB and dioxin-like PCB in indoor air of public buildings contaminated with different PCB sources – deriving toxicity equivalent concentrations from standard PCB congeners. Chemosphere 67:1746–1753

Hillwalker WE, Jepson PC, Anderson KA (2006) Selenium accumulation patterns in lotic and lentic aquatic systems. Sci Total Environ 366:367–379

Hori T, Nakagawa R, Tobiishi K, Lida T, Tsutumi T, Sasaki K, Toyoda M (2005) Effects of cooking on concerns of polychlorinated dibenzo-p-dioxins and related compounds in fish and meat. J Agric Food Chem 53:8820–8828

Hsueh YM, Hsu MK, Chiou HY, Yang MH, Huang CC, Chen CJ (2002) Urinary arsenic speciation in subjects with or without restriction from seafood dietary intake. Toxicol Lett 133:83–91

Huet M, Michel P, Averty B, Paulet YM (2003) Imposex-TBT. La pollution par les organostan-niques le long des côtes françaises, de la Manche et de l'Atlantique. Surveillance du Milieu Marin. Travaux du réseau national d'observation de la qualité du milieu marin: Bulletin RNO Edition 2003, Nantes (La Chapelle sur Erdre), 133p

INCA 1 (1999) Enquête individuelle et nationale sur la consommation alimentaire. Report. http://www.afssa.fr/index.htm

INCA 2 (2009) Enquête individuelle et nationale des consommations alimentaires 2 (INCA 2) 2006–2007. Report 228 p. http://www.afssa.fr/index.htm

INRS   (2010)   http://www.inrs.fr/inrs-pub/inrs01.nsf/IntranetObject-accesParReference/INRS-
    FR/$FILE/fset.html

IRSN (2010) http://www.irsn.fr/FR/Pages/home.aspx

Ineris (2010) http://www.ineris.fr/index.php

International chemical safety sheets (2010) http://training.itcilo.it/actrav_cdrom2/fr/osh/ic/
    alfamain.htm

JECFA (2010a) Joint FAO/WHO expert committee on food additives seventy-second meeting
    Rome, 16–25 February 2010 summary and conclusion. Issued 16th Mar 2010, 1–16

JECFA (2010b) Joint FAO/WHO expert committee on food additives seventy-third meeting
    Geneva, 8–17 June 2010 summary and conclusion. Issued 24th June 2010, 1–17

JEFCA (2001) Summary of the 57th meeting of the joint FAO/WHO expert committee on food
    additives. Rome, 5–14 June 2001

Journal Officiel de la République Française (JORF) (1999) Arrêté du 21 mai 1999 relatif
    au classement de salubrité et à la surveillance des zones de production et des zones de
    reparcage des coquillages vivants. Paris, pp 8508–8509. http://www.admi.net/jo/19990610/
    AGRM9901042A.html

James A, Claisse D, Marchand M (2006) Les normes de qualité environnementales (NQE), out-
    ils d'évaluation du bon état chimique. In: Ifremer (ed) RNO, Surveillance du Milieu Marin.
    Ifremer, Nantes, pp 20–26

Johnson MA, Paulet YM, Donval A, Le Pennec M (1996) Histology, histochemistry and enzyme
    biochemistry in the digestive system of the endosymbiont-bearing bivalve Loripes lucinalis
    (Lamarck). J Exp Mar Biol Ecol 197:15–38

Kaise T, Watanabe S, Itoh K (1985) The acute toxicity of arsenobetaine. Chemosphere 14:
    1327–1332

Kales SN, Huyck KL, Goldman RH (2006) Elevated urine arsenic: un-speciated results lead to
    unnecessary concern and further evaluations. J Anal Toxicol 30:80–85

Langston WJ (1983) The behaviour of arsenic in selected U.K. estuaries. Can J Fish Aquat Sci
    40:143–150

Leblanc JC, Guérin T, Noël L, Calamassi-Tran G, Volatier JL, Verger P (2005) Dietary exposure
    estimates of 18 elements from the 1st French total diet study. Food Add Contam: Part A 22:
    624–641

Leblanc JC, Volatier JL, Sirot V, Bemrah-Aouchia N (2006) CALIPSO: Fish and seafood con-
    sumption study and biomarker of exposure to trace elements, pollutants and omega 3. Report,
    162 p. http://www.afssa.fr/Documents/PASER-Ra-Calipso.pdf

Lewtas J (2007) Air pollution combustion emissions: characterization of causative agents and
    mechanisms associated with cancer, reproductive, and cardiovascular effects. Mutat Res
    636:95–133

Liber K, Culp JM, Kerrich R (2006) Importance de la spéciation de l'arsenic pour exprimer la
    toxicité de cet élément dans les organismes aquatiques: implications pour les recommandations
    pour la qualité des eaux du Canada, Santé Canada

Lim PE, Lee CK, Din Z (1998) The kinetics of bioaccumulation of zinc, copper, lead and cad-
    mium by oysters (Crassostrea iredalei and C. belcheri) under tropical field conditions. Sci
    Total Environ 216:147–157

Lorenzana RM, Yeow AY, Colman JT, Chappell LL, Choudhury H (2009) Arsenic in Seafood.
    Hum Ecol Risk Assess 15:185–200

Ma M, Le XC (1998) Effect of arsenosugar ingestion on urinary arsenic speciation. Clin Chem
    44:539–550

Manta DS, Angelone M, Bellanca A, Neri R, Sprovieri M (2002) Heavy metals in urban soils: a
    case study from the city of Palermo (Sicily), Italy. Sci Total Environ 300:229–243

Martoja R, Ballan-Dufrançais C, Jeantet AY, Gouzerth P, Amiard JC, Amiard-Triquet C, Berthet B,
    Baud JP (1988) Effets chimiques et cytologiques de la contamination expérimentale de l'huître
    Crassostrea gigas Thunberg par l'argent administré sous forme dissoute et par voie alimentaire.
    Can J Fish Aquat Sci 44:539–550

Merian E, Anke M, Ihnat M, Stoeppler M (2004) Elements and their compounds in the environment. Wiley-VCH, Weinheim

Metian M, Charbonnier L, Oberhaënsli F, Bustamante P, Jeffree R, Amiard JC, Warnau M. (2009) Assessment of metal, metalloid, and radionuclide bioaccessibility from mussels to human consumers, using centrifugation and simulated digestion methods coupled with radiotracer techniques. Ecotoxicol Environ Saf 72:1499–1502

Michel P (1993) L'arsenic en milieu marin. Biogéochimie et écotoxicologie. Repères Océan no. 4. Ifremer Edition, France, 62 p

Michel P, Averty B (1999) Contamination of French coastal waters by organotin compounds: 1997 Update. Mar Pollut Bull 38:268–275

Miramand P, Guary JC, Fowler SW (1980) Vanadium transfer in the mussel Mytilus galloprovincialis. Mar Biol 56:281–293

Mourgaud Y, Martinez E, Geffard A, Andral B, Stanisiere JY, Amiard JC (2002) Metallothionein concentration in the mussel Mytilus galloprovincialis as a biomarker of response to metal contamination: validation in the field. Biomarkers 7:479–490

Mozaffarian DMD, Rimm EB (2006) Fish intake, contaminants, and human health Evaluating the risks and benefits. JAMA 296:1885–1899

Munoz O, Devesa V, Suner MA, Velez D, Montoro R, Urieta I, Macho ML, Jalon M (2000) Total and inorganic arsenic in fresh and processed fish products. J Agric Food Chem 48: 4369–4376

Murray AP, Richardson BJ, Gibbs CF (1991) Bioconcentration factors for petroleum hydrocarbons, PAHs, LABs and biogenic hydrocarbons in the blue mussel. Mar Pollut Bull 22:595–603

Nakagawa R, Yumita Y, Hiromoto M (1997) Total mercury intake from fish and shellfish by Japanese people. Chemosphere 35:2909–2913

Narbonne JF, Michel X (1997) Systèmes de biotransformation chez les mollusques aquatiques. In: Lagadic L, Caquet T, Amiard JC, Ramade F (eds) Biomarqueurs en écotoxicologie. Aspects fondamentaux. Masson, Paris, pp 11–31

Noël L, Leblanc JC, Guérin T (2003) Determination of several elements in duplicate meals from catering establishment using closed vessel microwave digestion with inductively coupled plasma mass spectrometry detection: estimation of daily dietary intake. Food Addit Contam 20: 44–56

Noël L, Testu C, Chafey C, Pinte J, Velge P, Guérin T (in press) Contamination levels for lead, cadmium and mercury in big crustaceans: differences between white and brown meats. J Food Compos Anal

Noël L, Testu C, Chafey C, Velge P, Guérin T (2011) Contamination levels for lead, cadmium and mercury in marine gastropods, echinoderms and tunicates. Food Control 22:433–437

OSPAR (2007) Second periodic evaluation of progress towards the objective of the OSPAR radioactive substances strategy (JAMP product RA–2)

OSPAR (2008) The convention for the protection of the marine environment of the North-East Atlantic; 2007/2008 CEMP assessment: trends and concentration of selected hazardous substances in sediments and trends in TBT-specific biological effects. Available from http://www. ospar.org/documents/dbase/publications/p00378_2007-2008_CEMP_assessment.pdf

O'Connor TP (1998) Mussel watch results from 1986 to 1996. Mar Pollut Bull 37:14–19

Page DS, Dassanayake TM, Gilfillan ES (1995) Tissue distribution and depuration of tributyltin for field-exposed Mytilus edulis. Mar Environ Res 40:409–421

Pain S, Parant M (2003) Multixenobiotic defence mechanism (MXDM) in bivalves. Comptes Rendus Biol 326:659–672

Pilliére F, Conso F (2007) Biotox. In: INRS (ed) Guide biotoxicologique pour les médecins du travail. Paris, 252p. http://www.inrs.fr/inrs-pub/inrs01.nsf/IntranetObject-accesParReference/INRS-FR/$FILE/fset.html

Polikarpov GG (1960) Absorption of short-lived radioactivity by sea organisms. Priroda 49:105–7

Pradel J, Zettwoog P, Dellero N, Beutier D (2001) Le polonium 210, un repère naturel important en radioprotection. Radioprotection 36:401–416

Pruell RJ, Lake JL, Davis WR, Quinn JG (1986) Uptake and depuration of organic contaminants by blue mussels *(Mytilus edulis)* exposed to environmentally contaminated sediment. Mar Biol 91:497–507

Richardson SD, Ternes TA (2005) Water analysis: emerging contaminants and current issues. Anal Chem 77: 3807–3838

Roux N, Chiffoleau JF, Claisse D (2001) L'argent, le cobalt, le nickel et le vanadium dans les mollusques du littoral français. Surveillance du milieu marin. In: Ifremer (ed). Travaux RNO, 2001. Nantes (La Chapelle sur Erdre), pp 11–20

Saavedra Y, Fernández P, González A (2004) Determination of vanadium in mussels by electrothermal atomic absorption spectrometry without chemical modifiers. Anal Bioanal Chem 379:72–76

Sabbioni E, Fischbach M, Pozzi G, Pietra R, Gallorini M, Piette JL (1991) Cellular retention toxicity and carcinogenic potential of seafood arsenic. I. Lack of cytotoxicity and transforming activity of arsenobetaine in the BALB/3T3 cell line. Carcinogenesis 12:1287–1291

Samain JF, McCombie H (2008) Summer mortality of Pacific oyster *Crassostrea gigas* – The MOREST project

Sanders JG, Osman RW, Riedel GF (1989) Pathways of arsenic uptake and incorporation in estuarine phytoplankton and the filter-feeding invertebrates *Eurytemora affinis*, *Balanus improvisus* and *Crassostrea virginica*. Mar Biol 103:319–325

Sauvage C, Pépen JF, Lapégue S, Boudry P, Renault T (2009) Ostreid herpes virus 1 infection in families of the Pacific oyster, Crassostrea gigas, during a summer mortality outbreak: differences in viral DNA detection and quantification using real-time PCR. Virus Res 142: 181–187

Schaffner M, Bader HP, Scheidegger R (2009) Modeling the contribution of point sources and non-point sources to Thachin River water pollution. Sci Total Environ 407:4902–4915

Schoof RA, Yager JW (2007) Variation of total and speciated arsenic in commonly consumed fish and seafood. Hum Ecol Risk Assess 13:946–965

Schoof RA, Yost LJ, Eickhoff J, Crecelius EA, Cragin DW, Meacher DM, Menzel DB (1999) A market basket survey of inorganic arsenic in food. Food Chem Toxicol 37:839–846

SCOOP (2004) Report on tasks 3.2.11. Assessment of the dietary exposure to arsenic, cadmium, lead and mercury of the population of the EU Members States, pp 1–125. Available at http://ec.europa.eu/food/food/chemicalsafety/contaminants/scoop_3-2-11_heavy_metals_report_en.pdf

Sericano JL, Wade TL, Brooks JM (1996) Accumulation and depuration of organic contaminants by the American oyster (*Crassostrea virginica*). Sci Total Environ 179:149–160

Sharma VK, Sohn M (2009) Aquatic arsenic: toxicity, speciation, transformations, and remediation. Environ Int 35:743–759

Sirot V, Guérin T, Mauras Y, Garraud H, Volatier JL, Leblanc JC (2008) Methylmercury exposure assessment using dietary and biomarker data among frequent seafood consumers in France. Environ Res 107:30–38

Sirot V, Guérin T, Volatier J, Leblanc JC (2009) Dietary exposure and biomarkers of arsenic in consumers of fish and shellfish from France. Sci Total Environ 407:1875–1885

Soto M, Cajaraville MP, Marigómez I (1996) Tissue and cell distribution of copper, zinc and cadmium in the mussel, *Mytilus galloprovincialis*, determined by autometallography. Tissue Cell 28:557–568

Storelli MM, Ceci E, Marcotrigiano GO (1998) Comparison of total mercury, methylmercury, and selenium in muscle tissues and in the liver of *Stenella coeruleoalba* (Meyen) and *Caretta caretta* (Linnaeus). Bull Environ Contam Toxicol 61:541–547

Sukasem P, Tabucanon MS (1993) Monitoring heavy metals in the Gulf of Thailand using mussel watch approach. Sci Total Environ 139–140:297–305

SU-VI-MAX (2002) Portions alimentaires: manuel photos pour l'estimation des quantités. editor Polytechnica, Paris

Tripp BW, Farrington JW, Goldberg ED, Sericano J (1992) International mussel watch: the initial implementation phase. Mar Pollut Bull 24:371–373

Van Caneghem J, Block C, Van Brecht A, Wauters G, Vandecasteele C (2010) Mass balance for POPs in hazardous and municipal solid waste incinerators. Chemosphere 78:701–708

Vos G, Hovens JPC, Hagel P (1986) Chromium, nickel, copper, zinc, arsenic, selenium, cadmium, mercury and lead in dutch fishery products 1977–1984. Sci Total Environ 52:25–40

Walraven N, Laane RW (2009) Assessing the discharge of pharmaceuticals along the dutch coast of the North Sea. Rev Environ Contam Toxicol 199:1–18

Wei C, Li W, Zhang C, Van Hulle M, Cornelis R, Zhang X (2003) Safety evaluation of organoarsenical species in edible porphyra from the China Sea. J Agric Food Chem 51: 5176–5182

Yang RQ, Zhou QF, Jiang GB. (2006) Butyltin accumulation in the marine clam *Mya arenaria*: An evaluation of its suitability for monitoring butyltin pollution. Chemosphere 63:1–8

Lead Uptake, Toxicity, and Detoxification in Plants

Bertrand Pourrut, Muhammad Shahid, Camille Dumat, Peter Winterton, and Eric Pinelli

Contents

Bertrand Pourrut and Muhammad Shahid – equivalent first authors.

E. Pinelli (✉)
EcoLab (Laboratoire d'écologie fonctionnelle), INP-ENSAT, 31326 Castanet-Tolosan,
France; EcoLab (Laboratoire d'écologie fonctionnelle), UMR 5245 CNRS-INP-UPS, 31326
Castanet-Tolosan, France
e-mail: pinelli@ensat.fr

D.M. Whitacre (ed.), *Reviews of Environmental Contamination and Toxicology*,
Reviews of Environmental Contamination and Toxicology 213,
DOI 10.1007/978-1-4419-9860-6_4, © Springer Science+Business Media, LLC 2011

1 Introduction

Plants are the target of a wide range of pollutants that vary in concentration, speciation, and toxicity. Such pollutants mainly enter the plant system through the soil (Arshad et al. 2008) or via the atmosphere (Uzu et al. 2010). Among common pollutants that affect plants, lead is one of the most toxic and frequently encountered (Cecchi et al. 2008; Grover et al. 2010; Shahid et al. 2011). Lead continues to be used widely in many industrial processes and occurs as a contaminant in all environmental compartments (soils, water, the atmosphere, and living organisms). The prominence of environmental lead contamination results both from its persistence (Islam et al. 2008; Andra et al. 2009; Punamiya et al. 2010) and from its present and past numerous sources. These sources have included smelting, combustion of leaded gasoline, or applications of lead-contaminated media (sewage sludge and fertilizers) to land (Piotrowska et al. 2009; Gupta et al. 2009; Sammut et al. 2010; Grover et al. 2010). In 2009, production of recoverable lead from mining operations was 1690, 516, and 400 thousand metric tons by China, Australia, and the USA, respectively (USGS 2009).

Despite a long history of its beneficial use by humankind, lead has no known biological function in living organisms (Maestri et al. 2010) and is now recognized as a chemical of great concern in the new European REACH regulations (EC 1907/2006; Registration, Evaluation, Authorization, and Restriction of Chemicals). Moreover, lead was reported as being the second most hazardous substance, after arsenic, based on the frequency of occurrence, toxicity, and the potential for human exposure by the Agency for Toxic Substances and Disease Registry (ATSDR 2003). The transfer of lead from polluted soils to plants was therefore widely studied, especially in the context of food quality, use in phytoremediation, or in biotesting (Arshad et al. 2008; Uzu et al. 2009).

Lead is known to induce a broad range of toxic effects to living organism, including those that are morphological, physiological, and biochemical in origin. This metal impairs plant growth, root elongation, seed germination, seedling development, transpiration, chlorophyll production, lamellar organization in the chloroplast, and cell division (Sharma and Dubey 2005; Krzesłowska et al. 2009; Gupta et al. 2009, 2010; Maestri et al. 2010). However, the extent of these effects varies and depends on the lead concentration tested, the duration of exposure, the intensity of plant stress, the stage of plant development, and the particular organs studied. Plants have developed various methods for responding to toxic metal exposures. They have internal detoxification mechanisms to deal with metal toxicity that includes selective metal uptake, excretion, complexation by specific ligands, and compartmentalization (Gupta et al. 2009; Krzesłowska et al. 2010; Maestri et al. 2010; Sing et al. 2010; Jiang and Liu 2010).

The various responses of plants to lead exposure are often used as tools (bioindicators) in the context of environmental quality assessment. To develop tools that are relevant for ecotoxicological studies, it is essential to understand the mechanisms involved in plant uptake, transfer, and toxicity. This is especially true in selected research areas, such as choice of plant species, when polluted soils are under study (i.e., reduced transfer when studying vegetables or increased transfer when phytoextraction is desired). For example, legumes are considered more suitable to grow on contaminated soil than Umbelliferae, Liliaceae, Compositae, and Chenopodiaceae, because they take up reduced amounts of lead (Alexander et al. 2006). The reduced lead uptake by vegetables minimizes the threat of lead introduction to the food chain. In contrast, phytoextraction requires plants that can sequester excessive amounts of lead in their biomass without incurring damage to basic metabolic functions (Arshad et al. 2008; Zaier et al. 2010). *Pelargonium* (Arshad et al. 2008) and *Brassica napus* (Zaier et al. 2010) are characterized as Pb hyperaccumulators, and they can extract huge amounts of lead from contaminated soil without showing morphophytotoxicity symptoms. Indeed, these plants have efficient natural detoxification mechanism to alleviate lead toxicity. In this review, we propose to trace the relationship that exists between lead uptake, accumulation, translocation, and toxicity in plants.

2 Retention, Mobility, and Bioavailability of Lead in Soil

Lead occurs naturally in the earth's crust (Arias et al. 2010) and its natural levels remain below 50 mg kg^{-1} (Pais and Jones 2000). But, anthropogenic activities often modify the amount and nature of lead species present in soil. In soils, lead may occur as a free metal ion, complexed with inorganic constituents (e.g., HCO_3^-, CO_3^{2-}, SO_4^{2-}, and Cl^-), or may exist as organic ligands (e.g., amino acids, fulvic acids, and humic acids); alternatively lead may be adsorbed onto particle surfaces (e.g., Fe-oxides, biological material, organic matter, and clay particles) (Uzu et al. 2009; Tabelin and Igarashi 2009; Sammut et al. 2010; Vega et al. 2010). Anthropogenic-sourced lead generally accumulates primarily in the surface layer of soil, and its concentration decreases with depth (Cecchi et al. 2008). Because of its strong binding with organic and/or colloidal materials, it is believed that only small amounts of the lead in soil are soluble, and thereby available for plant uptake (Kopittke et al. 2008; Punamiya et al. 2010).

However, lead behavior in soil, in the context of species, solubility, mobility, and bioavailability, is largely controlled by complex interactions governed by many biogeochemical factors (Punamiya et al. 2010). These factors include pH (Kopittke et al. 2008; Lawal et al. 2010; Vega et al. 2010), redox conditions (Tabelin and Igarashi 2009), cation-exchange capacity (Vega et al. 2010), soil mineralogy (Dumat et al. 2006), biological and microbial conditions (Arias et al. 2010), amount of lead present (Bi et al. 2010; Cenkci et al. 2010; Lawal et al. 2010), organic and inorganic ligand levels (Padmavathiamma and Li 2010; Sammut et al. 2010; Shahid et al. 2011), competing cation levels (Kopittke et al. 2008; Komjarova and Blust 2009), and plant species involved (Kovalchuk et al. 2005; Bi et al. 2010; Liu et al. 2010). Such factors may act individually or in combination with each other and may alter the soil behavior of the lead present, as well as the rate of uptake by plants.

Lead bioavailability is strongly influenced by its speciation and, in particular, by the concentration of free lead ions present (Dumat et al. 2006; Uzu et al. 2009; Sammut et al. 2010; Shahid et al. 2011). This is because the most significant plant uptake route for many cationic metals (and especially for the free metal ion) is via the soil solution in dissolved form (Punamiya et al. 2010). Moreover, the free lead ion concentration in soils depends on the adsorption/desorption processes in which it participates (Vega et al. 2010).

3 Lead Behavior in Plants

3.1 Lead Uptake by Plants

With the exception of the special conditions that exist for plants cultivated near metal recycling industries (Uzu et al. 2010), the main pathway by which plants accumulate metals is through root uptake from soils (Sharma and Dubey 2005; Uzu et al. 2009). Part of the lead present in the soil solution is adsorbed onto the roots, and then becomes bound to carboxyl groups of mucilage uronic acid, or directly to the polysaccharides of the rhizoderm cell surface (Seregin and Ivanov 2001). Lead adsorption onto roots has been documented to occur in several plant species: *Vigna unguiculata* (Kopittke et al. 2007), *Festuca rubra* (Ginn et al. 2008), *Brassica juncea* (Meyers et al. 2008), *Lactuca sativa* (Uzu et al. 2009), and *Funaria hygrometrica* (Krzesłowska et al. 2009, 2010). Once adsorbed onto the rhizoderm roots surface, lead may enter the roots passively and follow translocating water streams. However, lead absorption is not uniform along plant roots as a lead concentration gradient from root apex can be observed (Tung and Temple 1996; Seregin et al. 2004). Indeed, the highest lead concentrations can be found in root apices, where root cells are young and have thin cell walls (with the exception of root cap cells) that facilitate root uptake (Tung and Temple 1996; Seregin et al. 2004). Moreover, the apical area is the area where rhizodermic pH is the lowest, which increases solubility of lead in the soil solution.

At the molecular level, the mechanism by which lead enters roots is still unknown. Lead may enter the roots through several pathways, and a particular pathway is through ionic channels. Although, lead uptake is a non-selective phenomenon, it nonetheless depends on the functioning of an H^+/ATPase pump to maintain a strong negative membrane potential in rhizoderm cells (Hirsch et al. 1998; Wang et al. 2007). Inhibition of lead absorption by calcium is well-known (Garland and Wilkins 1981; Kim et al. 2002) and is associated with competition between these two cations for calcium channels (Huang and Cunningam 1996). Several authors have demonstrated that Ca^{2+}-permeable channels are the main pathway by which lead enters roots (Wang et al. 2007; Pourrut et al. 2008). The use of transgenic plants has shown that lead can penetrate into roots through alternative non-selective pathways, such as cyclic nucleotide-gated ion channels (Arazi et al. 1999; Kohler et al. 1999) or via low-affinity cation transporters (Wojas et al. 2007).

Reduced uptake and translocation of lead to aerial plant parts of vegetables is considered to be beneficial in preventing lead from entering the food chain. However, reduced uptake and translocation of lead to aerial plant parts, when plants are used to remediate polluted soils, is a major problem. Indeed, soil remediation requires plants (hyperaccumulators) that can take high lead levels up and translocate it to aerial plant parts with no or minimal toxicity. The amount of lead that moves from soil to penetrate into plants can be measured by what is called "the transfer factor"; this factor is defined as the ratio that exists between the concentration of lead in the plant vs. the concentration of lead in the soil (Arshad et al. 2008; Bi et al. 2010; Liu et al. 2010). This transfer factor will be different for different plant species and will change as soil physical and chemical properties are altered (Arshad et al. 2008; Bi et al. 2010; Liu et al. 2010). Generally, plants having a transfer factor greater than 1 are categorized as hyperaccumulators, whereas those with transfer factor less than 1 are termed as non-accumulators of lead (Arshad et al. 2008).

3.2 Lead Accumulation in Plants

Once lead has penetrated into the root system, it may accumulate there or may be translocated to aerial plant parts. For most plant species, the majority of absorbed lead (approximately 95% or more) is accumulated in the roots, and only a small fraction is translocated to aerial plant parts, as has been reported in *Vicia faba, Pisum sativum*, and *Phaseolus vulgaris* (Piechalak et al. 2002; Małecka et al. 2008; Shahid et al. 2011), *V. unguiculata* (Kopittke et al. 2007), *Nicotiana tabacum,* (Gichner et al. 2008), *Lathyrus sativus* (Brunet et al. 2009), *Zea mays* (Gupta et al. 2009), *Avicennia marina* (Yan et al. 2010), non-accumulating *Sedum alfredii* (Gupta et al. 2010), and *Allium sativum* (Jiang and Liu 2010). Although many metals display the translocation restriction phenomenon mentioned above, this phenomenon is not common to all heavy metals. Notwithstanding, this phenomenon in plants is both specific and very intense for lead.

When entering the root, lead mainly moves by apoplast and follows water streams until it reaches the endodermis (Tanton and Crowdy 1971; Lane and Martin 1977). There are several reasons why the transport of lead from roots to aerial plant parts is limited. These reasons include immobilization by negatively charged pectins within the cell wall (Islam et al. 2007; Kopittke et al. 2007; Arias et al. 2010), precipitation of insoluble lead salts in intercellular spaces (Kopittke et al. 2007; Islam et al. 2007; Meyers et al. 2008; Małecka et al. 2008), accumulation in plasma membranes (Seregin et al. 2004; Islam et al. 2007; Jiang and Liu 2010), or sequestration in the vacuoles of rhizodermal and cortical cells (Seregin et al. 2004; Kopittke et al. 2007).

However, these reasons are not sufficient to explain the low rate of lead translocation from root to shoot. The endoderm, which acts as a physical barrier, plays an important role in this phenomenon. Indeed, following apoplastic transport, lead is blocked in the endodermis by the Casparian strip and must follow symplastic transport. In endodermis cells, the major part of lead is sequestered or excreted by plant detoxification systems (c.f. Section 5.2).

Several hyperaccumulator plant species, such as *Brassica pekinensis* and *Pelargonium*, are capable of translocating higher concentrations of lead to aerial plant parts, without incurring damage to their basic metabolic functions (Xiong et al. 2006; Liu et al. 2008; Arshad et al. 2008). A specific hyperaccumulator species can accumulate more than 1000 ppm lead (Maestri et al. 2010). Indeed, these plants exude substances from roots that dissolve metals in soil (Arshad et al. 2008) that increases uptake and translocation (by employing certain metal cation transporters/genes). Moreover, they can tolerate higher concentrations of lead ions because they have various detoxification mechanisms, which may include selective metal uptake, excretion, complexation by specific ligands, and compartmentalization.

In addition, translocation of lead to aerial plant parts increases in the presence of organic chelators like ethylenediaminetetraacetate (EDTA) (Liu et al. 2008; Zaier et al. 2010; Barrutia et al. 2010) or certain species of micro-organisms (Arias et al. 2010; Punamiya et al. 2010). Recently, Liu et al. (2010) reported that, in 30 *B. pekinensis* cultivars, increased soil lead levels also increased the percent translocation to aerial plant parts. High concentrations of lead are known to destroy the physical barrier formed by the Casparian strip.

Transportation of metals from plant roots to shoots requires movement through the xylem (Verbruggen et al. 2009) and, when it occurs, is probably driven by transpiration (Liao et al. 2006). Arias et al. (2010) demonstrated high lead deposition in xylem and phloem cells of mesquite plants by using X-ray mapping. After penetrating into the central cylinder of the stem, lead can again be transported via the apoplastic pathway. The lead is then translocated to leaf areas via vascular flow (Sharma and Dubey 2005; Krzesłowska et al. 2010). While passing through the xylem, lead can form complexes with amino or organic acids (Roelfsema and Hedrich 2005; Vadas and Ahner 2009; Maestri et al. 2010). However, lead may also be transferred in inorganic form, as is cadmium. To express the degree of lead translocation, some authors have used a translocation factor (lead in aerial parts/lead in roots) (Arshad et al. 2008; Uzu et al. 2009; Liu et al. 2010). When this factor is used, the numeric value is normally rather low, which indicates that lead has been sequestered in the roots (Uzu et al. 2009; Liu et al. 2010).

4 General Effects of Lead on Plants

4.1 Effects on Germination and Growth

When plants are exposed to lead, even at micromolar levels, adverse effects on germination and growth can occur (Kopittke et al. 2007). Germination is strongly inhibited by very low concentrations of Pb^{2+} (Tomulescu et al. 2004; Islam et al. 2007). Lead-induced inhibition of seed germination has been reported in *Hordeum vulgare*, *Elsholtzia argyi*, *Spartina alterniflora*, *Pinus halepensis*, *Oryza sativa*, and *Z. mays* (Tomulescu et al. 2004; Islam et al. 2007; Sengar et al. 2009). At higher concentrations, lead may speed up germination and simultaneously induce adverse affects on the length of radical and hypocotyl in *E. argyi* (Islam et al. 2007). Inhibition of germination may result from the interference of lead with protease and amylase enzymes (Sengar et al. 2009).

Lead exposure in plants also strongly limits the development and sprouting of seedlings (Dey et al. 2007; Gichner et al. 2008; Gopal and Rizvi 2008). At low concentrations, lead inhibits the growth of roots and aerial plant parts (Islam et al. 2007; Kopittke et al. 2007). This inhibition is stronger for the root, which may be correlated to its higher lead content (Liu et al. 2008). Lead toxicity may also cause swollen, bent, short and stubby roots that show an increased number of secondary roots per unit root length (Kopittke et al. 2007). Recently, Jiang and Liu (2010) reported mitochondrial swelling, loss of cristae, vacuolization of endoplasmic reticulum and dictyosomes, injured plasma membrane and deep colored nuclei, after 48–72 h of lead exposure to *A. sativum* roots. Arias et al. (2010) reported significantly inhibited root elongation in Mesquite (*Prosopis* sp.).

Plant biomass can also be restricted by high doses of lead exposure (Gopal and Rizvi 2008; Gichner et al. 2008; Islam et al. 2008; Piotrowska et al. 2009; Sing et al. 2010). Under severe lead toxicity stress, plants displayed obvious symptoms of growth inhibition, with fewer, smaller, and more brittle leaves having dark purplish abaxial surfaces (Islam et al. 2007; Gupta et al. 2009). Plant growth retardation from lead exposure may be attributed to nutrient metabolic disturbances (Kopittke et al. 2007; Gopal and Rizvi 2008) and disturbed photosynthesis (Islam et al. 2008). In most cases, the toxic effect of lead on plant growth is time and dose dependent (Dey et al. 2007; Gupta et al. 2009, 2010). However, the effect of low concentrations is not clearly established, and the observed growth inhibition is not necessarily correlated to a reduction in biomass (Kosobrukhov et al. 2004; Yan et al. 2010). Moreover, the effect of lead toxicity varies with plant species, i.e., hyperaccumulators naturally tolerate more lead toxicity than do sensitive plants (Arshad et al. 2008).

4.2 Effects on Proteins

Similar to what occurs with other heavy metals, lead interacts with cytoplasmic proteins. The effect of lead on the total concentration of protein is unclear, although high concentrations may decrease the protein pool (Chatterjee et al. 2004; Mishra et al. 2006; Garcia et al. 2006; Piotrowska et al. 2009). This quantitative decrease in total protein content is the result of several lead effects: acute oxidative stress of reactive oxygen species (ROS) (Piotrowska et al. 2009; Gupta et al. 2009), modification in gene expression (Kovalchuk et al. 2005), increased ribonuclease activity (Gopal and Rizvi 2008), protein utilization by plants for the purposes of lead detoxification (Gupta et al. 2009), and diminution of free amino acid content

(Gupta et al. 2009) that is correlated with a disturbance in nitrogen metabolism (Chatterjee et al. 2004). However, certain amino acids, like proline, increase under lead stress (Qureshi et al. 2007). Such proteins play a major role in the tolerance of the plant to lead. In contrast, low concentrations of lead increase total protein content (Mishra et al. 2006). This protein accumulation may defend the plant against lead stress (Gupta et al. 2010), particularly for proteins involved in cell redox maintenance. If true, then such proteins act in a way similar to how ascorbate functions or similar to how metals are sequestered by glutathione (GSH) or phytochelatins (PCs) (Brunet et al. 2009; Liu et al. 2009; Yadav 2010; Jiang and Liu 2010). In addition to a quantitative change, lead can affect the qualitative composition of cell proteins. The protein profile of root cells in bean seedlings was modified after lead exposure (Beltagi 2005). Such modification can be correlated to the change that occurs in the transcriptome profile of several enzymes including isocitrate lyase, cysteine proteinase *SAG12,* serine hydroxymethyltransferase, and arginine decarboxylase (Kovalchuk et al. 2005).

4.3 Water Status Effects

The disruption of plant water status after lead treatment has been addressed in many studies (Brunet et al. 2009). Results of such exposures show a decrease in transpiration, as well as reduction of the moisture content (Barcelo and Poschenrieder 1990; Patra et al. 2004). Reduced transpiration may result from reduced leaf surface area for transpiration that is caused by decreased leaf growth (Elibieta and Miroslawa 2005). However, some plant species that have high stomatal density are capable of coping with such effects (Kosobrukhov et al. 2004; Elibieta and Miroslawa 2005). Lead reduces plant cell wall plasticity, and thereby influences the cell turgor pressure. The decrease in concentrations of molecules that control cell turgor, such as sugars and amino acids, further accentuates the phenomenon of lead influence on turgor pressure (Barcelo and Poschenrieder 1990). The change in turgor pressure, particularly in the guard cells, interferes with stomatal opening and closing. To maintain cell turgor pressure, plants synthesize high concentrations of osmolytes, particularly proline under lead stress conditions (Qureshi et al. 2007).

Stomatal opening/closing is controlled by abscisic acid (ABA), a phytohormone (Roelfsema and Hedrich 2005). The presence of Pb^{2+} ions causes a large accumulation of ABA in roots and aerial plant parts (Parys et al. 1998; Atici et al. 2005; Cenkci et al. 2010), leading to stomatal closure (Mohan and Hosetti 1997). Stomatal closure strongly limits gas exchange with the atmosphere, and water losses by transpiration (Parys et al. 1998). According to Elibieta and Miroslawa (2005), the foliar respiration of plants is also reduced by lead exposure, because the deposition of a cuticle layer, for example, on *Glycine max* leaf surfaces, is affected. Moreover, a CO_2/O_2 imbalance in plants from lead-induced oxidative phosphorylation and respiratory disorders may also disrupt plant water status.

4.4 Mineral Nutrition Effects

Results from multiple studies demonstrate that nutrient uptake by plants is significantly affected by the presence of lead (Chatterjee et al. 2004; Sharma and Dubey 2005; Gopal and Rizvi 2008). Although data are insufficient to allow a definitive conclusion to be drawn, it is known that lead affects plant mineral uptake. It is also known that lead exposure decreases the concentration of divalent cations (Zn^{2+}, Mn^{2+}, Mg^{2+}, Ca^{2+}, and Fe^{2+}) in leaves of Z. *mays* (Seregin et al. 2004), *O. sativa* (Chatterjee et al. 2004), *Brassica oleracea* (Sinha et al. 2006), *Medicago sativa* (Lopez et al. 2007), *V. unguiculata* (Kopittke et al. 2007), and *Raphanus sativus* (Gopal and Rizvi 2008). But, it is not possible to conclude if this decrease results from blockage of root absorption, a decrease in translocation from roots to aerial plant parts, or a change in distribution of these elements in the plant. The effect of lead on mineral accumulation in aerial plant parts, in most cases, follows a common trend. In roots, the trend varies according to plant species or the intensity of the imposed stress (Lopez et al. 2007; Kopittke et al. 2007; Gopal and Rizvi 2008).

The decreased absorption of nutrient in the presence of lead may result from competition (e.g., those with atomic size similar to lead) or changes in physiological plant activities. According to Sharma and Dubey (2005), the strong interaction of K^+ ions with lead could result from their similar radii (Pb^{2+}: 1.29 Å and K^+: 1.33 Å): these two ions may compete for entry into the plant through the same potassium channels. Similarly, lead effects on K^+-ATPase and -SH groups of cell membrane proteins cause an efflux of K^+ from roots. However, lead does not cause nitrogen efflux. The general reduction in the concentration of inorganic nitrogen in all plant parts could be induced by the reduced activity of nitrate reductase, the rate-limiting enzyme in the nitrate assimilation process (Xiong et al. 2006; Sengar et al. 2009). Xiong et al. (2006) reported that lead exposure (4 and 8 mmol kg^{-1}) significantly decreased shoot nitrate content (70 and 80%), nitrate reductase activity (100 and 50%), and free amino acid content (81 and 82%) in *B. pekinensis*.

4.5 Photosynthesis Effects

Photosynthesis inhibition is a well-known symptom of lead toxicity (Xiong et al. 2006; Hu et al. 2007; Liu et al. 2008; Piotrowska et al. 2009; Sing et al. 2010; Cenkci et al. 2010). This inhibition is believed to result from the following indirect effects of lead rather than from a direct effect:

- distorted chloroplast ultrastructure from the affinity lead has for protein N and S ligands (Elibieta and Miroslawa 2005; Islam et al. 2007),
- decreased ferredoxin NADP$^+$ reductase and delta-aminolevulinic acid dehydratase (ALAD) activity at the origin of chlorophyll synthesis inhibition (Gupta et al. 2009; Cenkci et al. 2010),

- inhibition of plastoquinone and carotenoid synthesis (Kosobrukhov et al. 2004; Chen et al. 2007; Liu et al. 2008; Cenkci et al. 2010),
- obstruction of the electron transport system (Qufei et al. 2009),
- inadequate concentration of carbon dioxide via stomatal closure (Romanowska et al. 2002, 2005, 2006),
- impaired uptake of essential elements such as Mn and Fe (Chatterjee et al. 2004; Gopal and Rizvi 2008) and substitution of divalent cations by lead (Gupta et al. 2009; Cenkci et al. 2010),
- inhibition of Calvin cycle enzymatic catalysis (Mishra et al. 2006; Liu et al. 2008), and
- increased chlorophyllase activity (Liu et al. 2008).

However, these different effects vary by plant species. Generally, chlorophyll b is more sensitive than chlorophyll a (Xiong et al. 2006). The mechanism of chlorophyll breakdown into phytol, magnesium and the primary cleavage product of the porphyrin ring occur in four consecutive steps. This reaction is catalyzed by chlorophyllase, Mg-dechelatase, pheophorbide a oxygenase, and red chlorophyll catabolite reductase. Loss of the typical chlorophyll green color occurs only after cleavage of the porphyrin ring (Harpaz-Saad et al. 2007). The decrease observed in photosynthetic activity is often a more sensitive measure than is pigment content.

4.6 Respiration Effects

When exposed to lead, photosynthetic plants usually experience harmful effects on respiration and adenosine triphosphate (ATP) content. Unlike the photosynthetic activity, the effect of lead on respiratory activity has been little studied (Seregin and Ivanov 2001). All the studies carried out on respiratory activity deal with leaves, whereas the effect of the Pb^{2+} ions on the respiratory activity of roots remains unknown. Lead is reported to affect the activity of ribulose-bisphosphate carboxylase in C_3 plants that control CO_2 assimilation, without affecting oxygenase activity (Assche and Clijsters 1990). Therefore, it is quite possible that photosynthesis is significantly reduced without any effect on photorespiration being induced, thus increasing the relative ratio of photorespiration to photosynthesis. Parys et al. (1998) reported that the CO_2 concentration of P. sativum in leaves increased significantly after exposure to lead nitrate, most probably from the reduced photosynthetic and increased respiration activity. Romanowska et al. (2002) stressed that Pb^{2+}-induced increases in respiration resulted only from dark (mitochondrial) respiration, while photorespiration was unaffected. The stimulation of dark respiration by lead was observed in leaves or protoplasts of P. sativum and H. vulgare (Romanowska et al. 2002, 2005, 2006). Moreover, the stimulation of respiration was well correlated with increased production of ATP in mitochondria, resulting in the high energy demands of the plant to combat lead effects being met.

It has been also shown that divalent cations (e.g., Pb, Zn, Cd, Co, and Ni) can bind to mitochondrial membranes, disrupting the electron transport that could lead to decoupling of phosphorylation (Romanowska et al. 2002, 2006). An increase in the respiratory rate of 20–50% was observed by Romanowska et al. (2002) in the detached leaves of C_3 plants (*P. sativum* and *H. vulgare*) and C_4 plants (*Z. mays*), when they were exposed to 5 mM $Pb(NO_3)_2$ for 24 h. Glycine, succinate, and malate substrates were more fully oxidized in mitochondria, isolated from lead-treated *P. sativum* leaves, than in mitochondria from control leaves (Romanowska et al. 2002). Lead caused an increase in ATP content as well as an increase in the ATP/ADP ratio in *P. sativum* and *H. vulgare* leaves (Romanowska et al. 2005, 2006). Rapid fractionation of *H. vulgare* protoplasts, incubated under conditions of low and high CO_2, indicated that the increased ATP/ADP ratio in lead-treated leaves mainly resulted from the production of mitochondrial ATP. The activity of NAD^+-malate dehydrogenase in protoplasts of barley leaves treated with lead was threefold higher than that in protoplasts from control leaves (Romanowska et al. 2005). Lead also significantly inhibited Hill reaction activity in spinach chloroplasts, in addition to photophosphorylation; moreover, lead had a more conspicuous effect on cyclic photophosphorylation than on noncyclic photophosphorylation (Romanowska et al. 2008). Recently, Qufei and Fashui (2009) reported that the accumulation of Pb^{2+} in photosystem II resulted in damage to its secondary structure and induced decreased absorbance of visible light and inhibited energy transfer among amino acids. Moreover, Jiang and Liu (2010) reported mitochondrial swelling, loss of cristae, and vacuolization of endoplasmic reticulum and dictyosomes during a 48–72 h lead exposure in *A. sativum*.

4.7 Genotoxicity

The antimitotic effect of lead is one of its best known toxic effects on plants (Patra et al. 2004; Shahid et al. 2011). Indeed, Hammett (1928) demonstrated long ago that lead induces a dose-dependent decrease in mitotic activity in root cells of *Allium cepa*, which was later described in detail by Wierzbicka (1999) and Patra et al. (2004). In *V. faba* roots, lead shortened the mitotic stage and prolonged interphase, thus lengthening the cell cycle (Patra et al. 2004). The first step by which lead induces plant toxicity is the binding of the Pb^{2+} ion to cell membranes and to the cell wall. This produces rigidity in these components and reduces cell division. The second step is the disruption of microtubules that are essential for mitosis. Lead exposure induces disturbances in the G_2 and M stages of cell division that leads to the production of abnormal cells at the c-mitosis (colchicine-mitosis) stage. This phenomenon is thought to be accentuated by direct or indirect interactions of lead with the proteins involved in the cell cycle, such as cyclins. Cyclin activity is indirectly dependent on the concentration of GSH. The spindle activity disturbances caused by lead may be transient in some cases, returning the mitotic index to initial levels.

Unlike antimitotic mechanisms, the mechanisms by which lead causes genotoxicity are complex and not yet well understood. At a low concentration, lead did not induce a significant effect on mitosis, but did induce aberrations (chromosomal bridges during anaphase), loss of acentric fragments during meiosis, chromosomal fragmentation, and micronucleus formation (National Toxicology Program 2003; Patra et al. 2004; Cecchi et al. 2008; Marcato et al. 2009; Grover et al. 2010; Barbosa et al. 2010; Shahid et al. 2011). The induction of chromosomal aberrations by lead can be explained, in part, by its action of disrupting the microtubule network. Results of in vitro studies have demonstrated that lead creates breaks in single and double strands of DNA and thereby affects horizontal DNA–DNA or DNA–protein links (Rucińska et al. 2004; Gichner et al. 2008; Shahid et al. 2011).

Lead may enter the nucleus (Małecka et al. 2008) and bind directly to the DNA or indirectly to protein. After binding to DNA, lead disrupts DNA repair and replication mechanisms. Lead does not induce direct genotoxic effects until it becomes attached to naked DNA (Valverde et al. 2001). Lead can also affect replication by replacing the zinc in the Zn-finger pattern of the enzymes that intervene in DNA repair (Gastaldo et al. 2007). Recently, Cenksi et al. (2010) used a random amplified polymorphic DNA (RAPD) assay that amplifies random DNA fragments of genomic DNA, and they reported that genomic template stability was significantly affected by lead exposure in *Brassica rapa*.

4.8 Oxidative Stress and Lipid Peroxidation

ROS are produced during normal cell metabolism in the chloroplast, either as by-products of the reduction of molecular oxygen (O_2) or because of excitation in the presence of highly energized pigments. These ROS, such as superoxide radicals ($O_2^{\bullet-}$), hydroxyl radicals ($\bullet OH$), and hydrogen peroxide (H_2O_2), are also generated following exposure to certain environmental agents. The production of ROS in the cells of aerobic organisms, defined as oxidative stress, is a well-known feature of the toxicity of heavy metals, including lead (Pourrut et al. 2008; Liu et al. 2008; Grover et al. 2010; Yadav 2010; Singh et al. 2010). However, the degree to which this feature is important is dependent on the metal type, specific form of the metal, plant species, exposure time, etc. When ROS forms exhaust cellular antioxidant reserves, they can rapidly attack and oxidize all types of biomolecules, such as nucleic acids, proteins, and lipids (Reddy et al. 2005; Clemens 2006; Hu et al. 2007; Wang et al. 2007; Yadav 2010). Such attacks lead to irreparable metabolic dysfunction and cell death.

Lead causes marked changes in the lipid composition of different cell membranes (Liu et al. 2008; Piotrowska et al. 2009; Grover et al. 2010; Yan et al. 2010; Singh et al. 2010). The polyunsaturated fatty acids and their esters that are present in lipids show high susceptibility to ROS (Dey et al. 2007; Gupta et al. 2009). Indeed, ROS removes hydrogen from unsaturated fatty acids and forms lipid radicals and reactive aldehydes, ultimately causing distortion of the lipid bilayer (Mishra et al. 2006). Lead-induced changes in lipid composition and potassium ion leakage were reported in *Z. mays* (Malkowski et al. 2002). Lead ions are known to induce lipid peroxidation, decrease the level of saturated fatty acids, and increase the unsaturated fatty acid content of membranes in several plant species (Singh et al. 2010).

The oxidation of bis-allylic hydrogens on polyunsaturated fatty acids by ROS involves three distinct stages: initiation (formation of the lipid radical), progression (formation of lipid peroxyl radical by reaction between lipid radical and oxygen), and termination (formation of non-radical products after bimolecular interaction of lipid peroxyl radicals) (see details in the reviews of Gurer and Ercal 2000 and Bhattacharjee 2005). These lipid membrane changes cause the formation of abnormal cellular structures, such as alterations in the cell membrane (Dey et al. 2007; Islam et al. 2008; Gupta et al. 2009), organelles (e.g., mitochondria), peroxisomes (Małecka et al. 2008; Liu et al. 2008), or chloroplasts (Choudhury and Panda 2004; Elibieta and Miroslawa 2005; Hu et al. 2007).

5 Mechanisms of Lead Tolerance

Plants respond to noxious effects of lead in various ways, such as selective metal uptake, metal binding to the root surface, binding to the cell wall, and induction of antioxidants. There are several types of antioxidants to which plants may respond: non-protein thiol (NP-SH), cysteine, glutathione, ascorbic acid, proline, and antioxidant enzymes, such as superoxide dismutase (SOD), ascorbate peroxidase (APX), guaiacol peroxidase (GPX), catalase (CAT), and glutathione reductase (GR). However, the response varies with plant species, metal concentration, and exposure conditions.

5.1 Passive Mechanisms

Even when small amounts of lead penetrate root cell membranes, it interacts with cellular components and increases the thickness of cell walls (Krzesłowska et al. 2009, 2010). Pectin is a component of plant cell walls. Lead complexation with pectin carboxyl groups is regarded as the most important interaction by which plant cells can resist lead toxicity (Meyers et al. 2008; Jiang and Liu 2010). Krzesłowska et al. (2009) observed that binding of lead to JIM5-P (within the cell wall and its resultant thickening) acted as a physical barrier that restricted lead access to the plasma membrane in *F. hygrometrica* protonemata. However, later, these authors stated that lead bound to JIM5-P within the cell can be taken up or remobilized by endocytosis, together with this pectin epitope (Krzesłowska et al. 2010).

5.2 Inducible Mechanisms

Recently, several authors have reported the presence of transporter proteins among plant cells that play an important role in metal detoxification, by allowing the excretion of metal ions into extracellular spaces (Meyers et al. 2008; Vadas and Ahner 2009; Maestri et al. 2010). The human divalent metal transporter 1 (DMT1), expressed in yeast, has been shown to transport lead via a pH-dependent process (Bressler et al. 2004) in plants. Simultaneously, several ATP-binding cassette (ABC) carriers, such as AtATM3 or AtADPR12 at ATP-binding sites in *Arabidopsis*, were involved in resistance to lead (Kim et al. 2006; Cao et al. 2008). Although, suspected to act against lead, this detoxification mechanism has not yet been clearly confirmed. Transcriptome analysis has shown that the gene expression of these carriers is stimulated by lead (Liu et al. 2009).

Cellular sequestration is considered to be an important aspect of plant metal homeostasis and plant detoxification of heavy metals (Maestri et al. 2010). The lead, that could be bound by certain organic molecules (Piechalak et al. 2002; Vadas and Ahner 2009), is sequestered in several plant cell compartments: vacuoles (Małecka et al. 2008; Meyers et al. 2008), dictyosome vesicles (Malone et al. 1974), endoplasmic reticulum vesicles (Wierzbicka et al. 2007), or plasmatubules (Wierzbicka 1998).

Cysteine and glutathione (GSH) are known to be non-enzymatic antioxidants in plants. An increase in cysteine content, in response to lead toxicity, has been demonstrated in *Arabidopsis thaliana* (Liu et al. 2009). Glutathione protects plants from lead stress by quenching lead-induced ROS (Verbruggen et al. 2009; Liu et al. 2009). Moreover, as the substrate for phytochelatin (PC) biosynthesis, the glutathione-related proteins play an important role in heavy metal detoxification and homeostasis (Liu et al. 2009). Lead treatment can induce different GSH genes, including glutathione-synthetase, -peroxidase, and -reductase, and glutamylcysteine synthetase. Glutathione can also enhance accumulation of proline in stressed plants, a role that is associated with reducing damage to membranes and proteins (Liu et al. 2009). Gupta et al. (2010) reported the role of GSH in lead detoxification in *S. alfredii*, although this was accomplished without any induction of PC. This suggests that GSH may play an important role in detoxifying lead, under stress conditions where PCs are absent.

PCs and metallothioneins (MTs) are the best characterized metal-binding ligands in plant cells. These ligands belong to different classes of cysteine-rich heavy metal-binding protein molecules. PCs, the most frequently cited metal protective proteins in plants, are low molecular weight, metal-binding proteins that can form mercaptide bonds with various metals (Maestri et al. 2010) and play an important role in their detoxification in plants (Brunet et al. 2009; Liu et al. 2009; Gupta et al. 2010; Yadav 2010; Jiang and Liu 2010). These thiols are biologically active compounds, whose function is to prevent oxidative stress in plant cells (Verbruggen et al. 2009; Gupta et al. 2010). Their general structure is $(\gamma\text{-glutamyl-cys})_n$ Gly where $n = 2\text{--}11$, and they are synthesized by the action of γ-glutamylcysteine dipeptidyl transpeptidase (phytochelatin synthase; PCS) from GSH (Yadav 2010).

Lead is known to stimulate the production of PC and activate PCS (Mishra et al. 2006; Clemens 2006; Andra et al. 2009; Vadas and Ahner 2009; Sing et al. 2010). It has been proposed that in vivo, phytochelatins are involved in the cellular detoxification and accumulation of several metals, including lead, because of their ability to form stable metal–PC complexes (Clemens 2006; Yadav 2010). Phytochelatin sequesters soluble lead in the cytoplasm before transporting it to vacuoles and chloroplasts (Piechalak et al. 2002; Małecka et al. 2008; Jiang and Liu 2010), thus reducing the deleterious effect of Pb^{2+} in the cells. The mechanism regulating the passage of the lead–PC complex through the tonoplast is, however, not yet known. Gisbert et al. (2003) reported significantly increased uptake and tolerance to lead and Cd following the induction and over-expression of a wheat gene encoding for phytochelatin synthase (*TaPCS1*), in *Nicotiana glauca*.

5.3 Antioxidant Enzymes

To cope with the increased production of ROS and to avoid oxidative damage, plants have a system of antioxidant enzymes that scavenge the ROS that are present in different cell compartments (Brunet et al. 2009; Singh et al. 2010; Gupta et al. 2010). Lead-induced toxicity may inhibit the activity of these enzymes or may induce their synthesis (Table 1). However, lead-induced inhibition or induction of antioxidant enzymes is dependent on metal type, specific form of the metal, plant species type, and the duration/intensity of the treatment (Islam et al. 2008; Gupta et al. 2009; Singh et al. 2010).

Generally, lead inhibits enzymatic activities and, when this occurs, the values of the inactivation constant (K_i) range between 10^{-5} and 2×10^{-4} M (i.e., 50% of enzymatic activities are inhibited in this concentration range) (Seregin and Ivanov 2001). Enzyme inhibition results from the affinity lead has for -SH groups on the enzyme (Sharma and Dubey 2005; Gupta et al. 2009). This is true for more than 100 enzymes, including ribulose-1,5-bisphosphate carboxylase oxygenase (RuBisCO) and nitrate reductase. Inactivation results from a link at either the catalytic site or elsewhere on the protein and produces an altered tertiary structure. Lead can also produce the same effect by binding to protein-COOH groups (Gupta et al. 2009, 2010). Lead also interacts with metalloid enzymes. Indeed, lead can disrupt plant absorption of minerals that contain zinc, iron, manganese, etc., which are essential for these enzymes. Lead and other divalent cations also can substitute for these metals, and thereby inactivate enzymes, as occurs with ALAD (Gupta et al. 2009; Cenkci et al. 2010). The effect lead has on ROS constitutes another mechanism by which lead exposure affects protein behavior (Gupta et al. 2009, 2010).

Lead exposure is also known to stimulate the activities of certain enzymes (Table 1), but the mechanisms of action are, as yet, unclear. It has been proposed that lead activates certain enzymes by modulating gene expression or by restricting the activity of enzyme inhibitors (Seregin and Ivanov 2001). Indeed, antioxidant enzymes scavenge ROS, when they are produced in excess as a consequence of

Table 1 Lead-induced activation (↑) and reduction (↓) of enzymatic activities in different plant species

Plant species	↑ Enzyme activity	↓ Enzyme activity	References
Najas indica	SOD, GPX, APX, CAT, GR		Sing et al. (2010)
Sedum alfredii	SOD, APX		Gupta et al. (2010)
Zea mays	SOD, CAT, AsA		Gupta et al. (2009)
Lathyrus sativus	APX, GR, GST		Brunet et al. (2009)
Wolffia arrhiza	CAT, APX		Piotrowska et al. (2009)
Raphanus sativus	POD, ribonuclease	CAT	Gopal and Rizvi (2008)
Elsholtzia argyi	CAT	SOD, GPX	Islam et al. (2008)
Kandelia candel	SOD, POD, CAT		Zhang et al. (2007)
Bruguiera gymnorrhiza	SOD, POD, CAT		Zhang et al. (2007)
Cassia angustifolia	SOD, APX, GR, CAT		Qureshi et al. (2007)
Zea mays	SOD, POD, CAT		Wang et al. (2007)
Triticum aestivum	SOD, POX	CAT	Dey et al. (2007)
Potamogeton crispus	POD, SOD, CAT	SOD, CAT	Hu et al. (2007)
Ceratophyllum demersum	SOD, GPX, APX, CAT, GR	SOD, GPX, APX, CAT, GR	Mishra et al. (2006)
Helianthus annuus	GR	CAT	Garcia et al. (2006)
Macrotyloma uniflorum	SOD, CAT, POD, GR, GST		Reddy et al. (2005)
Cicer arietinum	SOD, CAT, POD, GR, GST		Reddy et al. (2005)
Taxithelium nepalense	APX, GPOX, CAT,		Choudhury and Panda (2004)
Oryza sativa	SOD, GPX, APX, CAT, GR	GR, CAT	Verma and Dubey (2003)

SOD, superoxide dismutase; APX, ascorbate peroxidase; GPX, guaiacol peroxidase; CAT, catalase; GR, glutathione reductase; AsA, ascorbic acid; GST, GSH S-transferase; GSH, glutathione; POD, peroxidase

metal toxicity. Superoxide dismutase, a metallo-enzyme present in various cell compartments, is considered to be the first defense against oxidative stress (Mishra et al. 2006). It catalyses the dismutation of two superoxide radicals to H_2O_2 and oxygen and thus maintains superoxide radicals at steady-state levels (Islam et al. 2008; Gupta et al. 2009). H_2O_2 is a very strong oxidant and requires quick removal to avoid oxidative toxicity; removal is achieved by the action of APX in the ascorbate-glutathione cycle or by GPX and CAT in the cytoplasm and in other cell compartments (Mishra et al. 2006). The role of GSH and glutathione reductase in the H_2O_2-scavenging mechanism in plant cells (Piechalak et al. 2002) is well established in the Halliwell–Asada enzyme pathway. Moreover, antioxidant enzymes may be activated from the increased concentration of their substrates, instead of direct interaction with lead (Islam et al. 2008).

6 Conclusions and Perspectives

Lead is a major inorganic global pollutant and numerous studies have revealed its biogeochemical behavior and impact on the biosphere. Based on these studies, as cited in this review, it is concluded as follows:

(1) Lead has been in use since antiquity, because of its many useful properties. The continued use of lead in many industrial processes has increased its concentration to toxic levels in all environmental compartments.

(2) Lead forms stable complexes with different compounds in soil and tends to be stored in the soil. The fate and behavior of lead in soil is affected by its form, solubility, mobility, and bioavailability and is controlled by many biogeochemical parameters, such as soil pH, redox conditions, cation-exchange capacity, soil mineralogy, biological and microbial conditions, amount and nature of organic and inorganic ligands present, and competing cations.

(3) Lead enters plants mainly through the roots via the apoplast pathway or calcium ion channels. Lead can also enter plants in small amounts through leaves. Once in the roots, lead tends to sequester in root cells. Only a limited amount of lead is translocated from roots to shoot tissues, because there are natural plant barriers in the root endodermis (e.g., Casparian strips).

(4) Lead has no biological function and induces various noxious effects inside plants. Excessive lead accumulation in plant tissue is toxic to most plants, leading to a decrease in seed germination, root elongation, decreased biomass, inhibition of chlorophyll biosynthesis, mineral nutrition and enzymatic reactions, as well as a number of other physiological effects. The intensity of these effects varies depending on the duration of exposure, stage of plant development, studied organ, and the concentration of lead used in the exposure assessment.

(5) Lead-induced production of ROS is the major cause of its toxicity. These free radicals disrupt the redox status of cells, causing oxidative stress and DNA damage through oxidation, and lead to irreparable metabolic dysfunction and cell death.

(6) Plants defend against lead toxicity through several avoidance or detoxification mechanisms. Plants resist lead entry into their cells via exclusion or they bind lead to their cell walls or other ligands. Plants combat lead-induced increased production of ROS by activating various antioxidant enzymes.

(7) The efficiency of detoxification mechanisms determines the final tolerance or sensitivity of plants to metal-induced stress. Plants that have efficient detoxification mechanisms are generally characterized as being hyperaccumulators. Such plants are useful in soil bioremediation for many metal types. Conversely, plants that do not efficiently cope with pollutants are sensitive to metal toxicity and are often used in risk assessment studies.

In this review, we raise several questions that need attention if our understanding of the biogeochemical behavior of lead in different environmental compartments is to be advanced. Lead is known to interfere directly or indirectly with the genetic material to induce ROS and modify (increase or decrease) the activities of certain enzymes in plants. These responses of plants to lead toxicity are often used as tools for risk assessment. However, the mechanisms of action underlying the noxious effects of lead in plants are still unknown.

Moreover, most field work performed on the effects of lead on plants is based almost exclusively on the total metal content in polluted soil, even though this is of little significance from an environmental point of view. Indeed, the potential effects of lead and other toxic elements in the environment depend on insights into their physico-chemical distribution, i.e., speciation. Therefore, if environmental scientists are to become better at predicting what the toxicity or environmental impact of lead may be, then additional research into the form of lead present is essential.

7 Summary

Lead has gained considerable attention as a persistent toxic pollutant of concern, partly because it has been prominent in the debate concerning the growing anthropogenic pressure on the environment. The purpose of this review is to describe how plants take lead up and to link such uptake to the ecotoxicity of lead in plants. Moreover, we address the mechanisms by which plants or plant systems detoxify lead.

Lead has many interesting physico-chemical properties that make it a very useful heavy metal. Indeed, lead has been used by people since the dawn of civilization. Industrialization, urbanization, mining, and many other anthropogenic activities have resulted in the redistribution of lead from the earth's crust to the soil and to the environment.

Lead forms various complexes with soil components, and only a small fraction of the lead present as these complexes in the soil solution are phytoavailable. Despite its lack of essential function in plants, lead is absorbed by them mainly through the roots from soil solution and thereby may enter the food chain. The absorption of lead by roots occurs via the apoplastic pathway or via Ca^{2+}-permeable channels. The behavior of lead in soil, and uptake by plants, is controlled by its speciation and by the soil pH, soil particle size, cation-exchange capacity, root surface area, root exudation, and degree of mycorrhizal transpiration. After uptake, lead primarily accumulates in root cells, because of the blockage by Casparian strips within the endodermis. Lead is also trapped by the negative charges that exist on roots' cell walls.

Excessive lead accumulation in plant tissue impairs various morphological, physiological, and biochemical functions in plants, either directly or indirectly, and induces a range of deleterious effects. It causes phytotoxicity by changing cell membrane permeability, by reacting with active groups of different enzymes involved in plant metabolism and by reacting with the phosphate groups of ADP or ATP, and by replacing essential ions. Lead toxicity causes inhibition of ATP production, lipid peroxidation, and DNA damage by over production of ROS. In addition, lead strongly inhibits seed germination, root elongation, seedling development, plant growth, transpiration, chlorophyll production, and water and protein content. The negative effects that lead has on plant vegetative growth mainly result from the following factors: distortion of chloroplast ultrastructure, obstructed electron transport, inhibition of Calvin cycle enzymes, impaired uptake of essential elements, such as Mg and Fe, and induced deficiency of CO_2 resulting from stomatal closure.

Under lead stress, plants possess several defense strategies to cope with lead toxicity. Such strategies include reduced uptake into the cell; sequestration of lead into vacuoles by the formation of complexes; binding of lead by phytochelatins, glutathione, and amino acids; and synthesis of osmolytes. In addition, activation of various antioxidants to combat increased production of lead-induced ROS constitutes a secondary defense system.

References

Alexander PD, Alloway BJ, Dourado AM (2006) Genotypic variations in the accumulation of Cd, Cu, Pb and Zn exhibited by six commonly grown vegetables. Environ Pollut 144: 736–745

Andra SS, Datta R, Sarkar D, Sarkar D, Saminathan SK, Mullens CP, Bach SB (2009) Analysis of phytochelatin complexes in the lead tolerant vetiver grass [*Vetiveria zizanioides* (L.)] using liquid chromatography and mass spectrometry. Environ Pollut 157(7):2173–2183

Arazi T, Sunkar R, Kaplan B, Fromm H (1999) A tobacco plasma membrane calmodulin-binding transporter confers Ni^{2+} tolerance and Pb^{2+} hypersensitivity in transgenic plants. Plant J 20:171–182

Arias JA, Peralta-Videa JR, Ellzey JT, Ren M, Viveros MN, Gardea-Torresdey JL (2010) Effects of *Glomus deserticola* inoculation on Prosopis: enhancing chromium and lead uptake and translocation as confirmed by X-ray mapping, ICP-OES and TEM techniques. Environ Exp Bot 68(2):139–148

Arshad M, Silvestre J, Pinelli E, Kallerhoff J, Kaemmerer M, Tarigo A, Shahid M, Guiresse M, Pradere P, Dumat C (2008) A field study of lead phytoextraction by various scented *Pelargonium* cultivars. Chemosphere 71(11):2187–2192

Assche F, Clijsters H (1990) Effects of metals on enzyme activity in plants. Plant Cell Environ 13(3):195–206

Atici Ö, Ağar G, Battal P (2005) Changes in phytohormone contents in chickpea seeds germinating under lead or zinc stress. Biol Plantarum 49(2):215–222

ATSDR (2003) Agency for Toxic Substances and Disease Registry. http://www.atsdr.cdc.gov/

Barbosa J, Cabral T, Ferreira D, Agnez-Lima L, Batistuzzo de Medeiros S (2010) Genotoxicity assessment in aquatic environment impacted by the presence of heavy metals. Ecotoxicol Environ Saf 73(3):320–325

Barceló J, Poschenrieder C (1990) Plant water relations as affected by heavy metal stress: a review. J Plant Nutr 13(1):1–37

Barrutia O, Garbisu C, Hernández-Allica J, García-Plazaola JI, Becerril JM (2010) Differences in EDTA-assisted metal phytoextraction between metallicolous and non-metallicolous accessions of *Rumex acetosa* L. Environ Pollut 158(5):1710–1715

Beltagi MS (2005) Phytotoxicity of lead (Pb) to SDS-PAGE protein profile in root nodules of faba bean (*Vicia faba* L.) plants. Pak J Biol Sci 8(5):687–690

Bhattacharjee S (2005) Reactive oxygen species and oxidative burst: roles in stress, senescence and signal transduction in plants. Curr Sci 89(7):1113–1121

Bi X, Ren L, Gong M, He Y, Wang L, Ma Z (2010) Transfer of cadmium and lead from soil to mangoes in an uncontaminated area, Hainan Island, China. Geoderma 155(1–2):115–120

Bressler JP, Olivi L, Cheong JH, Kim Y, Bannona D (2004) Divalent metal transporter 1 in lead and cadmium transport. Ann N Y Acad Sci 1012:142–152

Brunet J, Varrault G, Zuily-Fodil Y, Repellin A (2009) Accumulation of lead in the roots of grass pea (*Lathyrus sativus* L.) plants triggers systemic variation in gene expression in the shoots. Chemosphere 77(8):1113–1120

Cao X, Ma LQ, Singh SP, Zhou Q (2008) Phosphate-induced lead immobilization from different lead minerals in soils under varying pH conditions. Environ Pollut 152(1):184–192

Cecchi M, Dumat C, Alric A, Felix-Faure B, Pradere P, Guiresse M (2008) Multi-metal contamination of a calcic cambisol by fallout from a lead-recycling plant. Geoderma 144(1–2):287–298

Cenkci S, Cigerci IH, Yildiz M, Özay C, Bozdag A, Terzi H (2010) Lead contamination reduces chlorophyll biosynthesis and genomic template stability in *Brassica rapa* L. Environ Exp Bot 67(3):467–473

Chatterjee C, Dube BK, Sinha P, Srivastava P (2004) Detrimental effects of lead phytotoxicity on growth, yield, and metabolism of rice. Commun Soil Sci Plant Anal 35(1–2):255–265

Chen J, Zhu C, Li L, Sun Z, Pan X (2007) Effects of exogenous salicylic acid on growth and H_2O_2-metabolizing enzymes in rice seedlings under lead stress. J Environ Sci (China) 19(1):44–49

Choudhury S, Panda S (2004) Toxic effects, oxidative stress and ultrastructural changes in moss *Taxithelium Nepalense* (Schwaegr.) Broth. under chromium and lead phytotoxicity. Water Air Soil Pollut 167(1):73–90

Clemens S (2006) Evolution and function of phytochelatin synthases. J Plant Physiol 163(3): 319–332

Dey SK, Dey J, Patra S, Pothal D (2007) Changes in the antioxidative enzyme activities and lipid peroxidation in wheat seedlings exposed to cadmium and lead stress. Braz J Plant Physiol 19(1):53–60

Dumat C, Quenea K, Bermond A, Toinen S, Benedetti MF (2006) Study of the trace metal ion influence on the turnover of soil organic matter in cultivated contaminated soils. Environ Pollut 142(3):521–529

Elzbieta W, Miroslawa C (2005) Lead-induced histological and ultrastructural changes in the leaves of soybean (*Glycine max* (L.) Merr.). Soil Sci Plant Nutr 51(2):203–212

Garcia JS, Gratão PL, Azevedo RA, Arruda MAZ (2006) Metal contamination effects on sunflower (*Helianthus annuus* L.) growth and protein expression in leaves during development. J Agric Food Chem 54(22):8623–8630

Garland C, Wilkins D (1981) Effect of calcium on the uptake and toxicity of lead in *Hordeum vulgare* L. and *Festuca ovina* L. New Phytol 87(3):581–593

Gastaldo J, Viau M, Bencokova Z, Joubert A, Charvet A, Balosso J, Foray M (2007) Lead contamination results in late and slowly repairable DNA double-strand breaks and impacts upon the ATM-dependent signaling pathways. Toxicol Lett 173(3):201–214

Gichner T, Znidar I, Száková J (2008) Evaluation of DNA damage and mutagenicity induced by lead in tobacco plants. Mutat Res Genet Toxicol Environ Mutagen 652(2):186–190

Ginn BR, Szymanowski JS, Fein JB (2008) Metal and proton binding onto the roots of *Fescue rubra*. Chem Geol 253(3–4):130–135

Gisbert C, Ros R, De Haro A, Walker DJ, Pilar Bernal M, Serrano R, Navarro-Aviñó J (2003) A plant genetically modified that accumulates Pb is especially promising for phytoremediation. Biochem Biophys Res Commun 303(2):440–445

Gopal R, Rizvi AH (2008) Excess lead alters growth, metabolism and translocation of certain nutrients in radish. Chemosphere 70(9):1539–1544

Grover P, Rekhadevi P, Danadevi K, Vuyyuri S, Mahboob M, Rahman M (2010) Genotoxicity evaluation in workers occupationally exposed to lead. Int J Hyg Environ Health 213(2):99–106

Gupta D, Huang H, Yang X, Razafindrabe B, Inouhe M (2010) The detoxification of lead in *Sedum alfredii* H. is not related to phytochelatins but the glutathione. J Hazard Mater 177(1–3): 437–444

Gupta D, Nicoloso F, Schetinger M, Rossato L, Pereira L, Castro G, Srivastava S, Tripathi R (2009) Antioxidant defense mechanism in hydroponically grown *Zea mays* seedlings under moderate lead stress. J Hazard Mater 172(1):479–484

Gurer H, Ercal N (2000) Can antioxidants be beneficial in the treatment of lead poisoning? Free Radic Biol Med 29(10):927–945

Hammett FS (1928) Studies in the biology of metals. Protoplasma 5(1):535–542

Harpaz-Saad S, Azoulay T, Arazi T, Ben-Yaakov E, Mett A, Shiboleth YM, Hortensteiner S, Gidoni D, Gal-On A, Goldschmidt EE, Eyal Y (2007) Chlorophyllase is a rate-limiting enzyme in chlorophyll catabolism and is posttranslationally regulated. Plant Cell 19(3):1007–1022

Hirsch RE, Lewis BD, Spalding EP, Sussman MR (1998) A role for the AKT1 potassium channel in plant nutrition. Science 280(5365):918–921

Hu J, Shi G, Xu Q, Wang X, Yuan Q, Du K (2007) Effects of Pb^{2+} on the active oxygen-scavenging enzyme activities and ultrastructure in *Potamogeton crispus* leaves. Russ J Plant Physl 54(3):414–419

Huang JW, Cunningham SD (1996) Lead phytoextraction: species variation in lead uptake and translocation. New Phytol 134:75–84

Islam E, Liu D, Li T, Yang X, Jin X, Mahmood Q, Tian S, Li J (2008) Effect of Pb toxicity on leaf growth, physiology and ultrastructure in the two ecotypes of *Elsholtzia argyi*. J Hazard Mater 154(1–3):914–926

Islam E, Yang X, Li T, Liu D, Jin X, Meng F (2007) Effect of Pb toxicity on root morphology, physiology and ultrastructure in the two ecotypes of *Elsholtzia argyi*. J Hazard Mater 147(3):806–816

Jiang W, Liu D (2010) Pb-induced cellular defense system in the root meristematic cells of *Allium sativum* L. BMC Plant Biol 10:40–40

Kim D, Bovet L, Kushnir S, Noh EW, Martinoia E, Lee Y (2006) AtATM3 is involved in heavy metal resistance in Arabidopsis. Plant Physiol 140(3):922–932

Kim YY, Yang YY, Lee Y (2002) Pb and Cd uptake in rice roots. Physiol Plantarum 116:368–372

Kohler C, Merkle T, Neuhaus G (1999) Characterisation of a novel gene family of putative cyclic nucleotide- and calmodulin-regulated ion channels in *Arabidopsis thaliana*. Plant J 18(1):97–104

Komjarova I, Blust R (2009) Effect of Na, Ca and pH on simultaneous uptake of Cd, Cu, Ni, Pb, and Zn in the water flea *Daphnia magna* measured using stable isotopes. Aquat Toxicol 94(2):81–86

Kopittke PM, Asher CJ, Kopittke RA, Menzies NW (2007) Toxic effects of Pb^{2+} on growth of cowpea (*Vigna unguiculata*). Environ Pollut 150(2):280–287

Kopittke PM, Asher CJ, Kopittke RA, Menzies NW (2008) Prediction of Pb speciation in concentrated and dilute nutrient solutions. Environ Pollut 153(3):548–554

Kosobrukhov A, Knyazeva I, Mudrik V (2004) Plantago major plants responses to increase content of lead in soil: growth and photosynthesis. Plant Growth Regul 42(2):145–151

Kovalchuk I, Titov V, Hohn B, Kovalchuk O (2005) Transcriptome profiling reveals similarities and differences in plant responses to cadmium and lead. Mutat Res: Fundam Mol Mech Mutagen 570(2):149–161

Krzeslowska M, Lenartowska M, Mellerowicz EJ, Samardakiewicz S, Wozny A (2009) Pectinous cell wall thickenings formation–a response of moss protonemata cells to lead. Environ Exp Bot 65(1):119–131

Krzesłowska M, Lenartowska M, Samardakiewicz S, Bilski H, Woźny A (2010) Lead deposited in the cell wall of *Funaria hygrometrica* protonemata is not stable–a remobilization can occur. Environ Pollut 158(1):325–338

Lane SD, Martin ES (1977) A histochemical investigation of lead uptake in *Raphanus sativus*. New Phytol 79(2):281–286

Lawal O, Sanni A, Ajayi I, Rabiu O (2010) Equilibrium, thermodynamic and kinetic studies for the biosorption of aqueous lead(II) ions onto the seed husk of *Calophyllum inophyllum*. J Hazard Mater 177(1–3):829–835

Liao Y, Chien SC, Wang M, Shen Y, Hung P, Das B (2006) Effect of transpiration on Pb uptake by lettuce and on water soluble low molecular weight organic acids in rhizosphere. Chemosphere 65(2):343–351

Liu D, Li T, Jin X, Yang X, Islam E, Mahmood Q (2008) Lead induced changes in the growth and antioxidant metabolism of the lead accumulating and non-accumulating ecotypes of *Sedum alfredii*. J Integr Plant Biol 50(2):129–140

Liu T, Liu S, Guan H, Ma L, Chen Z, Gu H (2009) Transcriptional profiling of *Arabidopsis* seedlings in response to heavy metal lead (Pb). Environ Exp Bot 67(2):377–386

Liu X, Peng K, Wang A, Lian C, Shen Z (2010) Cadmium accumulation and distribution in populations of *Phytolacca americana* L. and the role of transpiration. Chemosphere 78(9):1136–1141

López ML, Peralta-Videa JR, Benitez T, Duarte-Gardea M, Gardea-Torresdey JL (2007) Effects of lead, EDTA, and IAA on nutrient uptake by alfalfa plants. J Plant Nutr 30(8): 1247–1261

Maestri E, Marmiroli M, Visioli G, Marmiroli N (2010) Metal tolerance and hyperaccumulation: costs and trade-offs between traits and environment. Environ Exp Bot 68(1):1–13

Malone C, Koeppe DE, Miller RJ (1974) Localization of lead accumulated by corn plants. Plant Physiol 53(3):388–394

Małecka A, Piechalak A, Morkunas I, Tomaszewska B (2008) Accumulation of lead in root cells of *Pisum sativum*. Acta Physiol Plant 30(5):629–637

Małkowski E, Kita A, Galas W, Karcz W, Kuperberg JM (2002) Lead distribution in corn seedlings (*Zea mays* L.) and its effect on growth and the concentrations of potassium and calcium. Plant Growth Regul 37(1):69–76

Marcato-Romain C, Guiresse M, Cecchi M, Cotelle S, Pinelli E (2009) New direct contact approach to evaluate soil genotoxicity using the *Vicia faba* micronucleus test. Chemosphere 77(3):345–350

Meyers DER, Auchterlonie GJ, Webb RI, Wood B (2008) Uptake and localisation of lead in the root system of *Brassica juncea*. Environ Pollut 153(2):323–332

Mishra S, Srivastava S, Tripathi R, Kumar R, Seth C, Gupta D (2006) Lead detoxification by coontail (*Ceratophyllum demersum* L.) involves induction of phytochelatins and antioxidant system in response to its accumulation. Chemosphere 65(6):1027–1039

Mohan BS, Hosetti BB (1997) Potential phytotoxicity of lead and cadmium to *Lemna minor* grown in sewage stabilization ponds. Environ Pollut 98(2):233–238

National Toxicology Program (2003) Report on carcinogens: background document for lead and lead compounds. Department of Health and Human Services, Research Triangle Park, NC

Padmavathiamma PK, Li LY (2010) Phytoavailability and fractionation of lead and manganese in a contaminated soil after application of three amendments. Bioresour Technol 101(14): 5667–5676

Pais I, Jones JB (2000) The handbook of trace elements. Saint Lucie Press, Boca Raton, FL, p 223

Parys E, Romanowska E, Siedlecka M, Poskuta J (1998) The effect of lead on photosynthesis and respiration in detached leaves and in mesophyll protoplasts of *Pisum sativum*. Acta Physiol Plant 20(3):313–322

Patra M, Bhowmik N, Bandopadhyay B, Sharma A (2004) Comparison of mercury, lead and arsenic with respect to genotoxic effects on plant systems and the development of genetic tolerance. Environ Exp Bot 52(3):199–223

Piechalak A, Tomaszewska B, Baralkiewicz D, Malecka A (2002) Accumulation and detoxification of lead ions in legumes. Phytochemistry 60(2):153–162

Piotrowska A, Bajguz A, Godlewska-Zylkiewicz B, Czerpak R, Kaminska M (2009) Jasmonic acid as modulator of lead toxicity in aquatic plant *Wolffia arrhiza* (Lemnaceae). Environ Exp Bot 66(3):507–513

Pourrut B, Perchet G, Silvestre J, Cecchi M, Guiresse M, Pinelli E (2008) Potential role of NADPH-oxidase in early steps of lead-induced oxidative burst in *Vicia faba* roots. J Plant Physiol 165(6):571–579

Punamiya P, Datta R, Sarkar D, Barber S, Patel M, Das P (2010) Symbiotic role of glomus mosseae in phytoextraction of lead in vetiver grass [*Chrysopogon zizanioides* (L.)]. J Hazard Mater 177(1–3):465–474

Qufei L, Fashui H (2009) Effects of Pb^{2+} on the Structure and Function of Photosystem II of *Spirodela polyrrhiza*. Biol Trace Elem Res 129(1):251–260

Qureshi M, Abdin M, Qadir S, Iqbal M (2007) Lead-induced oxidative stress and metabolic alterations in *Cassia angustifolia* Vahl. Biol Plantarum 51(1):121–128

Reddy AM, Kumar SG, Jyothsnakumari G, Thimmanaik S, Sudhakar C (2005) Lead induced changes in antioxidant metabolism of horsegram (*Macrotyloma uniflorum* (Lam.) Verdc.) and bengalgram (*Cicer arietinum* L.). Chemosphere 60(1):97–104

Roelfsema MRG, Hedrich R (2005) In the light of stomatal opening: new insights into 'the Watergate'. New Phytol 167(3):665–691

Romanowska E, Igamberdiev AU, Parys E, Gardeström P (2002) Stimulation of respiration by Pb^{2+} in detached leaves and mitochondria of C3 and C4 plants. Physiol Plant 116(2):148–154

Romanowska E, Pokorska B, Siedlecka M (2005) The effects of oligomycin on content of adenylates in mesophyll protoplasts, chloroplasts and mitochondria from Pb^{2+} treated pea and barley leaves. Acta Physiol Plant 27(1):29–36

Romanowska E, Wróblewska B, Drozak A, Siedlecka M (2006) High light intensity protects photosynthetic apparatus of pea plants against exposure to lead. Plant Physiol Biochem 44(5–6):387–394

Romanowska E, Wróblewska B, Drożak A, Zienkiewicz M, Siedlecka M (2008) Effect of Pb ions on superoxide dismutase and catalase activities in leaves of pea plants grown in high and low irradiance. Biol Plantarum 52(1):80–86

Rucińska R, Sobkowiak R, Gwóźdź EA (2004) Genotoxicity of lead in lupin root cells as evaluated by the comet assay. Cell Mol Biol Lett 9(3):519–528

Sammut M, Noack Y, Rose J, Hazemann J, Proux O, Depoux Ziebel M, Fiani E (2010) Speciation of Cd and Pb in dust emitted from sinter plant. Chemosphere 78(4):445–450

Sengar RS, Gautam M, Sengar RS, Sengar RS, Garg SK, Sengar K, Chaudhary R (2009) Lead stress effects on physiobiochemical activities of higher plants. Rev Environ Contam Toxicol 196:1–21

Seregin IV, Ivanov VB (2001) Physiological aspects of cadmium and lead toxic effects on higher plants. Russ J Plant Physiol 48(4):523–544

Seregin IV, Shpigun LK, Ivanov VB (2004) Distribution and toxic effects of cadmium and lead on maize roots. Russ J Plant Physiol 51(4):525–533

Shahid M, Pinelli E, Pourrut B, Silvestre J, Dumat C (2011) Lead-induced genotoxicity to *Vicia faba* L. roots in relation with metal cell uptake and initial speciation. Ecotoxicol Environ Saf 74(1):78–84

Sharma P, Dubey RS (2005) Lead toxicity in plants. Braz J Plant Physiol 17(1):35–52

Singh R, Tripathi RD, Dwivedi S, Kumar A, Trivedi PK, Chakrabarty D (2010) Lead bioaccumulation potential of an aquatic macrophyte *Najas indica* are related to antioxidant system. Bioresour Technol 101:3025–3032

Sinha P, Dube B, Srivastava P, Chatterjee C (2006) Alteration in uptake and translocation of essential nutrients in cabbage by excess lead. Chemosphere 65(4):651–656

Tabelin C, Igarashi T (2009) Mechanisms of arsenic and lead release from hydrothermally altered rock. J Hazard Mater 169(1–3):980–990

Tanton TW, Crowdy SH (1971) The distribution of lead chelate in the transpiration stream of higher plants. Pestic Sci 2(5):211–213

Tomulescu IM, Radoviciu EM, Merca VV, Tuduce AD (2004) Effect of copper, zinc and lead and their combinations on the germination capacity of two cereals. J Agric Sci 15

Tung G, Temple PJ (1996) Uptake and localization of lead in corn (*Zea mays* L.) seedlings, a study by histochemical and electron microscopy. Sci Total Environ 188(2–3):71–85

U.S. Geological Survey (2009) http://minerals.usgs.gov/minerals/pubs/commodity/lead/

Uzu G, Sobanska S, Aliouane Y, Pradere P, Dumat C (2009) Study of lead phytoavailability for atmospheric industrial micronic and sub-micronic particles in relation with lead speciation. Environ Pollut 157(4):1178–1185

Uzu G, Sobanska S, Sarret G, Munoz M, Dumat C (2010) Foliar lead uptake by lettuce exposed to atmospheric fallouts. Environ Sci Technol 44:1036–1042

Vadas TM, Ahner BA (2009) Cysteine- and glutathione-mediated uptake of lead and cadmium into *Zea mays* and *Brassica napus* roots. Environ Pollut 157(8–9):2558–2563

Valverde M, Trejo C, Rojas E (2001) Is the capacity of lead acetate and cadmium chloride to induce genotoxic damage due to direct DNA-metal interaction? Mutagenesis 16(3):265–270

Vega F, Andrade M, Covelo E (2010) Influence of soil properties on the sorption and retention of cadmium, copper and lead, separately and together, by 20 soil horizons: comparison of linear regression and tree regression analyses. J Hazard Mater 174(1–3):522–533

Verbruggen N, Hermans C, Schat H (2009) Molecular mechanisms of metal hyperaccumulation in plants. New Phytol 181:759–776

Verma S, Dubey RS (2003) Lead toxicity induces lipid peroxidation and alters the activities of antioxidant enzymes in growing rice plants. Plant Sci 164:645–655

Wang H, Shan X, Wen B, Owens G, Fang J, Zhang S (2007) Effect of indole-3-acetic acid on lead accumulation in maize (*Zea mays* L.) seedlings and the relevant antioxidant response. Environ Exp Bot 61(3):246–253

Wierzbicka M (1998) Lead in the apoplast of *Allium cepa* L. root tips–ultrastructural studies. Plant Sci 133(1):105–119

Wierzbicka M (1999) Comparison of lead tolerance in *Allium cepa* with other plant species. Environ Pollut 104(1):41–52

Wierzbicka MH, Przedpełska E, Ruzik R, Ouerdane L, Połeć-Pawlak K, Jarosz M, Szpunar J, Szakiel A (2007) Comparison of the toxicity and distribution of cadmium and lead in plant cells. Protoplasma 231(1):99–111

Wojas S, Ruszczynska A, Bulska E, Wojciechowski M, Antosiewicz DM (2007) Ca^{2+}-dependent plant response to Pb^{2+} is regulated by LCT1. Environ Pollut 147(3):584–592

Xiong Z, Zhao F, Li M (2006) Lead toxicity in *Brassica pekinensis* Rupr.: effect on nitrate assimilation and growth. Environ Toxicol 21(2):147–153

Yadav S (2010) Heavy metals toxicity in plants: an overview on the role of glutathione and phytochelatins in heavy metal stress tolerance of plants. S Afr J Bot 76(2):167–179

Yan ZZ, Ke L, Tam NFY (2010) Lead stress in seedlings of *Avicennia marina*, a common mangrove species in South China, with and without cotyledons. Aquat Bot 92(2):112–118

Zaier H, Ghnaya T, Ben Rejeb K, Lakhdar A, Rejeb S, Jemal F (2010) Effects of EDTA on phytoextraction of heavy metals (Zn, Mn and Pb) from sludge-amended soil with *Brassica napus*. Bioresour Technol 101(11):3978–3983

Zhang F, Wang Y, Lou Z, Dong J (2007) Effect of heavy metal stress on antioxidative enzymes and lipid peroxidation in leaves and roots of two mangrove plant seedlings (*Kandelia candel* and *Bruguiera gymnorrhiza*). Chemosphere 67(1):44–50

Before the Curtain Falls: Endocrine-Active Pesticides – A German Contamination Legacy

Ulrike Schulte-Oehlmann, Jörg Oehlmann, and Florian Keil

Contents

1 Introduction

As a result of the European Parliament approving a new EU pesticide regulation (1107/2009/EC replacing directive 91/414/EEC) and a directive on the sustainable use of pesticides (2009/128/EC), in October 2009, various active ingredients are likely to be banned for use as pesticides. The use of pesticides that are carcinogenic, mutagenic, and toxic to reproduction, or that have endocrine-disrupting properties, shall no longer be authorized for use. Active ingredients that are persistent, bioaccumulative and toxic (PBT), or very persistent and very bioaccumulative

U. Schulte-Oehlmann (✉)
Department of Aquatic Ecotoxicology, Institute of Ecology, Evolution and Diversity,
Goethe University Frankfurt am Main, 60323 Frankfurt am Main, Germany
e-mail: schulte-oehlmann@bio.uni-frankfurt.de

D.M. Whitacre (ed.), *Reviews of Environmental Contamination and Toxicology*,
Reviews of Environmental Contamination and Toxicology 213,
DOI 10.1007/978-1-4419-9860-6_5, © Springer Science+Business Media, LLC 2011

(vPvB) shall be phased out as well. The decision-making process for setting test criteria for endocrine-disrupting pesticides is pending and is planned to be finalized by 2013 (EU 2009a). The new regulation becomes effective in June 2011. According to directive 2009/128/EC, all member states are required to adopt National Action Plans for reducing the human health and environmental risks of pesticide use. The protection of the aquatic environment and drinking water supplies from pesticides, and the obligation to undertake corresponding control measures, was particularly highlighted.

According to the Statistical Office of the European Union, the overall pesticide consumption of all 25 EU member states was 219,771 t/a (annum) in 2003; in Germany alone, the consumption was 23,240 t/a (equating to 10.6% of the total) (Eurostat 2007). Germany's consumption in the EU is exceeded by only three countries: France with 61,753 t/a (28.1%), Spain with 31,815 t/a (14.5%), and Italy with 30,828 t/a (14.0%). Moreover, if the pesticide consumption of the United Kingdom is also considered (14,920 t/a or 6.8%), approximately 75% of the total pesticide consumption of the EU is allotted to these five member states (Eurostat 2007).

In 2003, fungicides played the most important role in the EU's total pesticide consumption (49%), followed by herbicides (38%), insecticides/molluscicides and others (10%), as well as plant growth regulators (3%) (Eurostat 2007). Mancozeb and inorganic sulfur represented the most frequently applied active ingredients among fungicides. Among herbicides, farmers most frequently used glyphosate and isoproturon, whereas pest insects mainly succumbed from use of chlorpyrifos and parathion-methyl (these have been excluded from the EU list of approved active ingredients since 2003). The highest application rates of pesticides occurred in European viniculture (average dosage used by crop 21.4 kg active ingredient/ha) and market gardening (average dosage used by crop 61.7 kg active ingredient/ha).

In 2010, roughly 1,200 pesticides were authorized under the German Plant Protection Act (BVL 2010), comprising a total of approximately 250 different active ingredients. The highest consumption rate of pesticides in Germany is allotted to cultivation of grain (12,000 t in 2003; application rates of about 2 kg active ingredient/ha). German users employ more herbicides (54%) than fungicides (34%), compared to the European average. In 2003, isoproturon, metazachlor, mancozeb, and inorganic sulfur (chemical most commonly used to protect grapes against powdery mildew) represented Germany's most frequently used pesticides. Among field crops, the German application rate of pesticides in potato cultivation (6 kg active ingredient/ha) is comparatively high but is exceeded by application rates in the fruit- (about 20 kg active ingredient/ha) and wine-growing (30 kg active ingredient/ha) sectors (Eurostat 2007).

2 Endocrine-Active Pesticide Ingredients

In recent years, several authors and expert panels have attempted to evaluate the endocrine-disrupting properties of pesticides. For this review, we have evaluated the metadata from nine pertinent lists and databanks to determine which of the

250 active ingredients currently used in Germany are suspected to have endocrine-disrupting properties (BKH Consulting Engineers & TNO Nutrition and Food Research 2000; BMELV 2009; DHI Water & Environment 2007; FOOTPRINT 2010; Kemikalieinspektionen 2008; McKinlay et al. 2008; Neumeister and Reuter 2008; Pesticides Safety Directorate 2008; RPS-BKH Consulting Engineers et al. 2002). The result is that 41 chemicals (16.9% of all substances used in Germany) appear in at least one of the lists or databanks evaluated. Ioxynil, mancozeb, and maneb were cited most frequently and were included on seven of the nine lists or databases. Bifenthrin (status on Annex I – approved pesticides under directive 91/414/EEC– is pending but has been resubmitted), deltamethrin, iprodione, metiram, and metribuzin were indicated as endocrine disrupters in five lists and the following active ingredients were included on up to four of the nine lists: 2,4-D, carbendazim, dimethoate, epoxiconazole, metconazole, picloram, prochloraz, tebuconazole, thiram, and triadimenol. The remaining 23 chemicals were referred to on these lists only once or twice. In summary, the azoles (triazoles and imidazoles; 13 substances or 31.7%), the dithiocarbamates/carbamates (five substances or 12.2%), and the pyrethroids (five substances or 12.2%) were rated remarkably often as having endocrine-disrupting properties.

However, we emphasize that, in this review, we do not intend to challenge or affirm whether or not the classification of a substance as an endocrine disrupter is reasonable. We are distinctly aware that a substance classification scheme will not be conclusive until the European Commission decides on corresponding test criteria (see above). Therefore, in this chapter, our intent is to give an account of the current state of the discussion regarding contamination of the environment by potentially endocrine-disrupting components of pesticides.

3 Routes of Pesticide Emission into the Environment

In Germany, approximately 80% of all pesticides are employed in agriculture and the remaining 20% are used for bib-agricultural applications to public areas (e.g., roadsides and parkways), shopping malls, and residential areas (Bavarian Environment Agency 2008). Active ingredients are known to be emitted via diffuse (spray drift, evaporation, runoff, leaching, erosion, and drainage) or point source discharges (courtyard drains, industrial discharge, municipal sewage plant discharge, etc.) into the environment.

In 1993 and 1994, the Federal Environment Agency modeled the distribution of such discharges into German surface waters of the 42 pesticide-active ingredients most frequently used in agriculture (UBA 2000). The total emission was calculated to be 30 t/a, which is equivalent to 0.1% of the total amount of pesticides used during that period (margin of uncertainty 10–70 t/a). The most important losses among diffuse discharges (~15 t/a or 50% of the total; margin of uncertainty 2–40 t/a) were runoff (9 t/a; 30%), spray drift (3.5 t/a; 12%), and drainage (1.5 t/a; 5%). Additionally, courtyard drains contributed 10 t/a (33%; margin of uncertainty

7–22 t/a) to water pollution by pesticides. Isolated releases of direct industrial discharges (only river Rhine area) were calculated to be less than 4 t/a (13%). Discharges from municipal wastewater treatment plants were not considered.

3.1 Spray Drift

Up to 10% of the active ingredient concentrations measured in treated crops can be detected in adjacent untreated plants (Bavarian Environment Agency 2008). During spring spraying applications to fruit crops, more than 10% of the applied pesticides are lost by spray drift, whereas this value in grain and vegetable crops is only 1% (Bavarian Environment Agency 2008). Carter (2000) evaluated field monitoring data and calculated spray-drift deposition levels for arable crop treatments, and reported depositions of between 0.3 and 3.5% at a 1-m distance from the handling area. Bach et al. (2005) used DRIPS (drainage, runoff, and spray-drift input of pesticides in surface waters) modeling to calculate a total loss of 38 kg of active ingredient via spray drift, following arable crop treatments in Germany. This equates to approximately 0.0003% of the total amount of applied active ingredients or to 14,053 t of the cumulative value of 59 active chemicals applied to arable cropland in 2000. Spray-drift losses from vineyard and fruit-growing areas have been reported to be 120 and 3,100 kg/a, respectively (Huber et al. 2000; Bach et al. 2001).

3.2 Runoff

The capacity for soil to absorb water or retain pesticides depends on the characteristics of the soil to which the pesticides are applied. Some soils retain little water or pesticides, whereas others may retain considerable amounts. Therefore, in addition to runoff, soluble pesticides and those bound to particulates may be horizontally translocated across the application areas (surface runoff). Neumann et al. (2002) observed measurable field runoff when precipitation exceeded 10 mm/day. Torrential rain events excluded, Carter (2000) indicated that the pesticide loss rate originating from farmland was generally less than 0.05%. Bach et al. (2005), however, estimated the runoff rates of 59 active ingredients for field crop treatment to be 14.9 t/a, which equates to 0.11% of the total amount (14,053 t) of these 59 substances applied in Germany during the year 2000.

According to Neumann et al. (2002), the application rate and octanol/water partition coefficients (P_{OW}) of active ingredients determine the level of measurable pesticide load by which different routes of entry (surface runoff, courtyard drains, storm water sewers, emergency overflows, or final effluents) contribute to the contamination of small bodies of running water. Generally P_{OW} values are negatively correlated with measured pesticide loads. This finding is traced back to the tendency of lipophilic substances to bind to particulate matter. However, for the different routes of entry analyzed by these authors, the P_{OW} as a determinant for pesticide load was confirmed only for surface runoff.

3.3 Volatilization

On the basis of a literature review comprising 28 European studies from 10 EU countries, Dubus et al. (2000) reported that 50% of 99 chemically analyzed pesticide-active ingredients (including isomers and metabolites) were found in rainwater. Measured concentrations were generally below 100 ng/L. Occasionally, maximum concentrations in the low microgram per liter range were detected. According to Carter (2000), the loss of pesticides via evaporation for most products did not exceed 20% of the amount applied. However, for extremely volatile substances, up to 90% of the applied amount may evaporate. In contrast, Huber (1998) indicated volatilization loss of pesticides in Germany to be only 50 kg/a (equivalent to approximately 0.0002% of the total German pesticide consumption). Carter (2000) concluded that, compared to the total unwanted contamination of the environment from agricultural pesticides, contamination from atmospheric deposition originating from rain, snow, and fog is marginal.

3.4 Leaching and Drainage

Leaching is the main process by which pesticides reach groundwater. Substance loss through lateral and vertical infiltration into groundwater typically constitutes less than an average of 1% of the amounts applied, and in more exceptional cases up to 5% (Carter 2000). Based on drainage water measurements, Bach et al. (2005) calculated a loss of 185 kg of pesticides resulting from 2003 field crop treatments in Germany. This corresponds to 0.0013% of 14,053 t of active ingredients used for arable crop production (based on the sales volume for the top 59 active ingredients used in agriculture in Germany in the year 2000). Carter (2000) uses a value that is 760-fold higher as a basis and predicts pesticide loss from field drainage to be up to 1% (equivalent to 140 t used in German field crop protection).

3.5 Point Sources

Direct discharges may account for up to 90% of a water body's pesticide load (Bavarian Environment Agency 2008). Direct discharges include those from industrial sources, from courtyards or other hard-surfaced areas (railroad tracks, sealed private, and public grounds), from which pesticides reach water bodies either directly or via sewage treatment plant (STP) effluents. Several authors (Bach 1999; Seel et al. 1996; Fischer et al. 1998; Müller et al. 2002) have assumed that municipal STP may contribute between 65 and 95% of the pesticide load that reaches small bodies of running water. Bach et al. (2005) determined that, depending on the substance, up to 100% of a single chemical contamination incident can be traced back to point source emissions for river catchments. Over a 3-year period, Altmayer et al. (2003) investigated 24-h mixed samples of two STPs that received multiple discharges from vineyards contaminated by pesticides commonly used in viniculture. Occasionally, active ingredient daily loads of up to 100 g were detected.

Bach et al. (2001) reported that in Germany, agricultural point sources can contribute up to 18 t/a to the total pesticide contamination of the aquatic environment. In other studies, it has been determined that single farms released between 5 and 80 g/year of active ingredients, during the periods measured (Bach et al. 2005). Neumann et al. (2002) investigated the catchment basins of two small creeks (Nette and Pletschbach) in the lower Rhine area. They focused on direct and indirect discharges originating from courtyard drains (3 of 25 adjacent farmsteads); one effluent stream included an emergency overflow and one a storm sewer that drained surface runoff from a farmed area (7 of 20 adjacent fields). Analyses were made of two insecticides, five fungicides, and thirteen herbicides during the main pesticide application period between April and mid-July 1998. The aqueous phase of the surface runoff samples contained 19 of 20 analyzed active ingredients, adding up to a total chemical load of 66.2 g, within the sampling period. Courtyard water samples contained 17 of 20 ingredients and an average amount of 24 g of all measured substances. The total substance load was 604 g, within the sampling period. Rainwater samples had residues of 20 analyzed chemicals. The estimated total substance load for rainwater was 18.5 g. No fungicides or insecticides were detected, but 11 and 12 herbicides were present in the emergency overflow and final sewer samples, respectively. The total active ingredient load measured in the final sewer effluent was 3,070 g, and the emergency overflow load was 925 g.

4 Ground and Drinking Water Contamination

According to BMG (German Federal Ministry of Health) and UBA (German Federal Environment Agency) (2006, 2008), German drinking water is of good to very good quality. Both reports refer to communications made by the 16 German states regarding 2,706 (in 2006) and 2,624 (in 2008) drinking water analyses provided by water supply companies. Only drinking water suppliers that attained an average daily flow rate of more than 1,000 m^3 or those serving more than 5,000 people were considered. An amount equal to 74% of the raw waters investigated, during the reporting period 2005–2007, originated from groundwater (76.1% during 2002–2004), 15.5% from surface water (13.3% during 2002–2004), and 10% from other sources (10.5% during 2002–2004), such as bank filtration and artificially enriched groundwater. During the reporting period 2002–2004, the EU reference values of 0.1 µg/L for a single active ingredient, and 0.5 µg/L for the sum of measured substances (EU drinking water directive 98/83/EC), were exceeded only in 1–2% of all samples taken (during 2005–2007, this value was <1%). From these analyses, local health authorities did observe long-term deviations from allowed maximum concentrations for pesticides and their metabolites in drinking water, predominantly for atrazine, bromacil, desethylatrazine (atrazine metabolite), 2,6-dichlorobenzamide (dichlobenil metabolite), and *N,N*-dimethylsulfamide (DMS, tolyfluanide metabolite).

In 2006, Sturm et al. (2007) carried out a study on the pesticide contamination of ground- and surface waters. The authors consulted surveys of 477 members from the German Technical and Scientific Association for Gas and Water (DVGW), excerpts of the groundwater data bank from Baden-Wuerttemberg and results of a federal state monitoring program for groundwater by the LAWA (Working group of Federal States on Water issues). Results were that 182 participating DVGW member waterworks (38% of all waterworks considered) reported positive findings of active ingredients or their metabolites that exceeded the limit of detection. However, these values did not necessarily exceed the EU drinking water reference value of 0.1 µg/L (for a single substance). Of all findings, 65% referred to groundwater, 31.0% to surface water, 4% to bank-filtered water, or artificially enriched groundwater, and 0.2% to other water sources. The number of analyzed parameters and frequency of sampling varied among the sampled waterworks, which is why identifying representative analyses (even those calculated from single-substance average concentrations) was impossible. In total, positive findings of 100 different substances were reported. Of these, 43% were approved substances (according to EU directive 91/414/EEC), 50% were prohibited, and 7% represented metabolites. The drinking water reference value of 0.1 µg/L was exceeded for 82% of all positive findings. Active ingredients found most often (listed more than 120 times) were atrazine and desethylatrazine, followed by diuron, simazine, isoproturon, and 2,6-dichlorobenzamide (number of times mentioned, 40–60). The number of times that bentazone, mecoprop, deisopropylatrazine, and terbuthylazine was mentioned ranged from 20 to 40. Hexazinone, propazine, metaxon (MCPA), chlortolurone, desethylbutylazine, and metazachlor were reported as having been detected 10–20 times by the waterworks. Five to ten positive findings occurred for the following metabolites and active ingredients: AMPA (metabolite of glyphosate), dichlorprop, glyphosate, metolachlor, ethidimuron, 1,2-dichloropropane, 2,4-D, bromoxynil, flufenacet, lenacil, metalaxyl, methabenzthiazuron, terbutryn (banned since 2003 as an active ingredient in herbicides but still approved in biocides), and ethofumesate.

Waterworks reported a total of 60 positive findings for active ingredients in groundwater. At the time of inquiry (2006), 10% of these substances were metabolites, 44% approved, and 47% no longer approved by EU pesticide regulators. For 41 substances or their metabolites (68% of all active ingredients and 6.8% of all positive findings), concentrations exceeding 0.1 µg/L (drinking water reference value) were detected. In some groundwater samples, maximum concentrations exceeded 1 µg/L (Sturm and Kiefer 2009).

A nationwide comparison of groundwater monitoring data was performed by LAWA for the periods 1990–1996 (LAWA 1998) and 1996–2000 (LAWA 2004). The studies made clear that, over the course of the preceding decade, pesticide contamination of groundwater remained unchanged. The comparison also indicated that regulatory inspections were largely consistent across the German Länder and confirmed the above-mentioned results of the waterworks. This nationwide data evaluation also demonstrated that atrazine and its metabolites, as well as bentazone, bromacil, diuron, and simazine, were most frequently detected in groundwater.

Kiefer and Sturm (2008) used their results as an opportunity to compile a list of pesticide-active ingredients and their metabolites that have "very high" relevance for water pollution control measures (Table 1). Eleven of 43 substances have been indicated as potential endocrine disrupters. Of these, only bromoxynil and metribuzin are still approved, according to the EU pesticide directive (Table 1).

5 Surface Water Contamination

During the main annual pesticide application period, waters from the rivers Danube, Main, Regnitz, and Altmühl in Bavaria, as well as small streaming waters, are frequently analyzed for residues of 100–150 active ingredients. According to Wagener and Schuster (2007), in small Bavarian streams, both the number and concentrations of pesticide-active ingredients are higher than those found in large watercourses.

Atrazine and its metabolites, terbutryn and metolachlor, were the endocrine-disrupting pesticides most often detected (Wagener and Schuster 2007) in both small and large streams. An average metolachlor maximum concentration of 0.29 μg/L (average value from 22 sampling stations spanning eight analytical studies) was detected in small streams. The LAWA environmental quality standard (EQS) requires the protection of aquatic biocoenosis at values of <0.2 μg metolachlor/L. The average maximum concentration of atrazine measured in small Bavarian watercourses met the LAWA and ICPR (International Commission for the Protection of the Rhine) target of 0.1 μg atrazine/L (drinking water and biocoenosis protection) and EU EQS of 0.6 μg atrazine/L (surface waters). In large Bavarian rivers, values were even lower. Details on the 90 percentile concentrations have not yet been provided but will become available. For terbutryn, an EQS was not defined by LAWA or any other responsible commission.

In 2002, the most important findings that concerned residues of pesticides with potentially endocrine-disrupting properties in the rivers Danube, Neckar, Rhine, Enz, Jagst, Kocher, and Tauber related to substances that no longer have authorization under applicable EU pesticide regulations. Such pesticides include alachlor, atrazine, diazinon, simazine, and terbutryn. However, active ingredients that are still approved in the EU, such as diuron, penconazole, pendimethalin, and propiconazole (LUBW – Environment Agency Baden-Wuerttemberg 2004), were also detected. For atrazine, the 90 percentile reference values of LAWA, ICPR, and IKSE (International Commission for Protection of the Elbe River) were not exceeded in any of the sampled rivers during the period of investigation. Nevertheless, some authors (Moltman et al. 2007) have proposed lower atrazine and simazine reference values (0.01 μg/L), based on ecotoxicological effect data. The 90 percentile

Table 1 Plant protection products (substances and metabolites) that have been detected most frequently by water suppliers in Germany and substances with "very high relevance" for prevention of water pollution (data of Sturm and Kiefer (2007) and Kiefer and Sturm (2008) upgraded and extended)

Substance name	Approved under directive 91/414/EEC (status October 2010)	Rank (according to water supplier's positive findings)	Rank (according to positive findings in groundwater)	Rank (according to positive findings in surface water)	Detected in groundwater	Detected in surface water	Detected in drinking water	Potential endocrine disrupter (according to lists/databanks consulted)
1,2-Dichloropropane	No	n.i.	n.i.	n.i.	X	X		No
2,6-Dichlorobenzamide	Dichlobenil metabolite	6	3	18	X	X	X	n.i.
Ametryn	No	n.i.	n.i.	n.i.		X	X	No
AMPA	Glyphosate metabolite	18	21	12	X	X		n.i.
Atrazine	No	2	2	2	X	X	X	Yes
Bentazone	Yes	7	7	9	X	X	X	No
Bromacil	No	8	6	n.i.	X	X	X	No
Bromoxynil	Yes	n.i.	n.i.	n.i.	X	X	X	Yes
Carbofuran	No	n.i.	n.i.	n.i.		X	X	Yes
Chloridazon	Yes	n.i.	n.i.	n.i.	X	X	X	No
Chlorotoluron	Yes	15	15	19	X	X	X	No
Cyanazine	No	n.i.	n.i.	n.i.	X	X		Yes
Desethylatrazine	Atrazine metabolite	1	1	7	X	X	X	n.i.
Desethylterbuthylazine	Terbuthylazine metabolite	16	14	17	X	X	X	n.i.
Deisopropylatrazine	Atrazine/simazine metabolite	10	9	19	X	X	X	n.i.
Desmetryn	No	n.i.	n.i.	n.i.		X	X	No
Diphenylchloridazon	Chloridazon metabolite	n.i.	n.i.	n.i.	X	X	X	n.i.
Dichlorprop	No	20	19	16	X	X	X	No
Dinoterb	No	n.i.	n.i.	n.i.	X	X	X	No
Diuron	Yes	3	5	1	X	X	X	Yes

Table 1 (continued)

Substance name	Approved under directive 91/414/EEC (status October 2010)	Rank (according to water supplier's positive findings)	Rank (according to positive findings in groundwater)	Rank (according to positive findings in surface water)	Detected in groundwater	Detected in surface water	Detected in drinking water	Potential endocrine disrupter (according to lists/databanks consulted)
Ethidimuron	No	n.i.	n.i.	n.i.	X	X	X	No
Fenpropimorph	Yes	n.i.	n.i.	n.i.	X		X	No
Flufenacet	Yes	n.i.	n.i.	n.i.		X		No
Glyphosate	Yes	20	24	10	X	X		Vague
Hexazinone	No	12	11	19	X	X	X	No
Isoproturon	Yes	5	8	3	X	X	X	No
Lenacil	Yes	n.i.	n.i.	n.i.	X	X		No
Metaxon	Yes	14	24	5	X	X		No
Mecoprop	Yes	9	10	8	X	X	X	No
Metalaxyl	No	n.i.	n.i.	n.i.	X	X		No
Metazachlor	Yes	17	19	13	X	X	X	No
Methabenzthiazuron	No	n.i.	n.i.	n.i.	X	X	X	Yes
Metolachlor	No	19	22	11	X	X	X	Yes
Metribuzin	Yes	n.i.	n.i.	n.i.	X	X	X	Yes
Metsulfuron-methyl	Yes	n.i.	n.i.	n.i.	X	X	X	No
N,N-Dimethylsulfamide	Tolyfluanide metabolite	n.i.	n.i.	n.i.	X	X	X	n.i.
Prochloraz	No	n.i.	n.i.	n.i.		X	X	Yes
Prometryn	No	n.i.	n.i.	n.i.	X	X	X	Yes
Propazine	No	13	12	19	X	X	X	Yes
Sebuthylazine	Not listed	n.i.	n.i.	n.i.	X		X	No
Simazine	No	4	4	4	X	X	X	Yes
Terbuthylazine	No	11	13	6	X	X	X	No
Terbutryn	No	n.i.	n.i.	n.i.	X	X	X	Yes

Abbreviations: X, positive finding ($\geq$ limit of detection); n.i., no information available of the range of diuron residues found in the river Kocher was 0.12 µg/L; this value was considerably higher than the EQS for aquatic organisms recommended by LAWA (0.05 µg/L) and ICPR (0.006 µg/L). The concentrations of diazinon found in the sampled water bodies exceeded the calculated EQS of 0.003 µg/L proposed by Moltmann et al. (2007). For the other detected substances, no reference values have been provided by the river commissions for waters that are near the surface. However, active ingredients have often been detected in such waters at a concentration range that exceeded the detection limit.

Between 1985 and 2003, the Environment Agency of Rhineland-Palatinate (LUWG) carried out a monitoring program on organic trace elements in running waters. In total, analyses were conducted for 144 pesticides, biocides, and 13 pesticide metabolites (LUWG 2006). From this monitoring program, a total of 48,948 measurements were made of water samples from the rivers Rhine, Moselle, Lahn, Nahe, Saar, and Selz and from water samples taken from selected smaller watercourses. Among those analytes covered were 22 fungicides, 73 herbicides, 56 insecticides, 2 nematicides, and 1 growth regulator.

In total, 157 pesticide-active ingredients were addressed in the monitoring study. Among these, 90 (57.3%) were not detectable and 67 (42.7%) had concentrations above the detection limit. A 50% quota (i.e., 50% of all measured concentrations were higher than the detection limit for at least one sampling station over a period of 1 year) existed for 29 active ingredients. Tebuconazole concentrations in the rivers Nahe, Moselle, and Selz exceeded the detection limit (0.03–0.05 µg tebuconazole/L). Water samples from the rivers Rhine, Lahn, and Saar were negative for tebuconazole residues. In 2001, the river Selz displayed annual average values of between 0.075 and 0.53 µg tebuconazole/L (maximum value 4.7 µg/L). Quality criteria for tebuconazole concentrations in surface waters are, unfortunately, currently not specified.

The pesticides that are potentially endocrine active, such as atrazine (and its metabolites), simazine (both now banned), diuron, and metazachlor, were similarly detected and had values above the 50% quota. For example, diuron (with a detection limit of <0.1 µg/L) was consistently detected in 10 of the water bodies (44% of the samples contained concentrations above the detection limit) for which analyses were performed. Annual mean values for diuron were between 0.025 and 0.326 µg/L. A maximum value of 1.5 µg/L was measured in the river Moselle more than one decade ago, in 1995. The ICPR EQS for aquatic biocoenosis of 0.006 µg diuron/L was, thus, frequently exceeded. Although application restrictions were placed on diuron, no concentration decrease was observed (LUWG 2006). Between 1988 and 2003, metazachlor (with a detection limit of 0.01–0.12 µg/L) was analyzed for in 24 watercourses and was detected in the rivers Rhine, Selz, Nahe, Moselle, and Saar and the brook Schwarzbach. The 50% quota for metazachlor was exceeded in the rivers Rhine (1988, 1992) and Selz (1997). The annual mean residue value detected for this herbicide was 0.032 µg/L. The maximum value was 0.39 µg/L and was measured in the river Selz in 1997.

Almost 70% of all atrazine residue values measured between 1988 and 2003 exceeded the limit of detection (0.01–0.55 µg/L), and these were mostly observed after the application of atrazine was banned in 1991. A failure to detect atrazine occurred only in eight of 24 water bodies, and the 50% quota was exceeded in 12 of the 24. The annual average residue values for atrazine ranged from 0.013 to 0.354 μg/L and were therefore above the EQS of 0.01 μg/L that was recommended by Moltmann et al. (2007). The highest atrazine residue detected was 2.1 μg/L and was recorded in 1995 in the river Moselle. In addition, 39% of all simazine concentrations detected exceeded the limit of detection (0.01–0.1 μg/L). This substance was present in 20 of 24 running water bodies. Annual average residue values for simazine (0.012–0.355 μg/L; maximum value 1.54 μg/L in river Selz in 1998) were comparable to those for atrazine and therefore probably exceeded the LAWA and ICPR EQS values. For some water bodies a gradual decline of the residue levels for atrazine and simazine was observed.

Among 56 insecticides analyzed, only a few potential endocrine disrupters appeared to exceed the limit of detection (parathion-methyl/-ethyl, α-endosulfan, and dimethoate). Only lindane (gamma-HCH) exceeded the detection limit of 0.001–0.02 μg/L, in approximately half (46%) of all measurements performed, in 22 streams between 1985 and 2001. At sampling stations in which the 50% quotes were exceeded, the yearly average values were in the range of 0.01–0.37 μg/L and thus were partially above the EQS of 0.066 μg/L proposed by Moltmann et al. (2007). The proposed EU Water Framework Directive (WFD) EQS of 0.02 μg/L for lindane (EU 2006), however, was not achieved at the end of the 1980s and the beginning of the 1990s in the rivers Lahn, Moselle, Saar, and Wiesbach. A maximum value of 0.12 μg lindane/L was measured in the Moselle in 1987. Although lindane was sporadically detected until 2003, the ban on lindane across all EU countries since 2001 turned out to be effective, because, in general, measured concentrations have been declining (LUWG 2006).

Until 2003, several streams were intensively monitored. Results of those pesticide-active ingredients that exceeded the 50% quota in sampled rivers were as follows:

- in the Rhine, 14 of 113 (equivalent to 12%) active ingredients;
- in the Moselle (10 of 89) and Saar (7 of 60), the total equating to ~11%;
- in the Nahe (14 of 73), approximately 19%;
- in the Selz (18 of 91), approximately 20%; and
- in the Lahn (7 of 27), approximately 26%.

The contamination patterns among the sampled rivers differed greatly. Over a period of 4 years, there were exceedances of the 50% quota for the following pesticides: dichlorprop, 2,4-D, MCPA, diuron, isoproturon, bentazone, chloridazon, and lindane. The rivers involved and the number of exceedances were as follows: Moselle (17), Saar (14), Rhine (9), Lahn (8), Selz (6), and Nahe (4).

In 2006, the pesticide monitoring network of the federal state Brandenburg addressed a total of 23 active ingredients and metabolites, spanning 17 sampling stations at the rivers Elbe, Odra, Neisse, Havel, Spree, Dahme, Nuthe, Rhin, Dosse, Stepenitz, Odra-Spree Canal, and Schwarze Elster. Positive findings occurred for the following 15 active ingredients: atrazine, 2,4-D, DDT, DDE, DDD, dichlorprop, α/β-endosulfan, lindane isomers, MCPA, mecoprop, glyphosate, isoproturon, metolachlor, pendimethalin, and terbuthylazine. These active ingredients were detected most frequently in the river Odra (nine substances or 39%), as well as in the rivers Havel and Schwarze Elster (seven substances or 30%). In 2006, violations of quality standards for pesticide residues were observed only for the herbicides dichlorprop and mecoprop. In contrast to the results of the preceding years, no positive findings were reported for aldrin, bentazone, chloridazon, chlortolurone, ethephon, or metazachlor (MUGV – Brandenburg State Office for the Environment 2007).

For the river Elbe, annual average residue values for 2006 were compared to the EQS of the EU WFD for several pesticides (2,4-D, aldrin, ametryn, atrazine, dichlorprop, dieldrin, dimethoate, diuron, endrin, hexazinone, isoproturon, MCPA, mecoprop, metazachlor, metolachlor, parathion-methyl, prometryn, simazine, and terbuthylazine). Results showed that all sampling stations retained good water quality. Also, tailwater areas of major tributaries, such as the rivers Schwarze Elster, Mulde, Saale, and Havel, were not significantly charged with residues (results were generally less than the limit of detection). At only one site in 2007 was there an exception; a water body near Schmilka displayed a p,p'-DDT annual average residue value that was twice the EU EQS standard (ARGE Elbe 2008a, b).

Investigations into contamination of Hessian streams were carried out either between 2004 and 2005, or between 2007 and 2009 by the Hessian State Office for Environment and Geology (HLUG 2010; data available at www.hlug.de/medien/wasser/wasser_psm/index.htm). In 2004 and 2005, a total of 122 sampling stations were examined six times annually (four samplings in spring and two in autumn) for 94 active ingredients and their metabolites. In a subsequent monitoring program (2007–2009), one-third of these stations were sampled. Herein, 74 substances were investigated and results compared with the WFD standards. In summary, HLUG found that surface waters situated in areas that have a distinct agricultural utilization profile and wastewater-loaded streams are characterized by extensive pesticide contamination. The Hessian Minister for Environment, Agriculture, and Consumer Protection, Wilhelm Dietzel, compiled a list addressing maximum pesticide residue concentrations of 21 active ingredients and their metabolites measured in Hesse during the 2004/2005 sampling campaign at 25 sampling stations (Hessian State Parliament 2006). Of these, primarily bentazone, isoproturon, diuron, dichlorprop, MCPA, mecoprop, and metamitron were detected.

The development program "Rhine 2020" of the ICPR aims at improving the water quality of the river Rhine. As part of the program, a list of contaminants relevant to the river Rhine (considering the OSPAR and WFD priority substances) is kept, along with the corresponding quality standards (ICPR 2007). Measured values for the banned chemicals aldrin, azinphos-ethyl, dieldrin, DDT, endrin, α-, β-, δ-HCH, isodrin, malathion, and simazine were in line with the established quality standards. Concentrations that were either considerably higher or partially above those standards were detected for alachlor, atrazine, azinphos-methyl, chlorfenvinphos, dichlorprop, dichlorvos, endosulfan, fenitrothion, fenthion, lindane, parathion-methyl/-ethyl, and trifluralin (banned substances according to

EU legislation). Approved substances (some of them presumably endocrine active), such as bentazone, chlorpyrifos, dimethoate, diuron, and metaxon, similarly exceeded quality standards.

In 2001, 23 pesticide-active ingredients were analyzed for in the river Danube. Among the detected residues, both atrazine and desethylatrazine were found to have average concentrations of 0.05 μg/L (ICPDR 2002). Some residue levels appeared to exceed the ICPR and LAWA EQS for atrazine (0.1 μg/L) in the tributaries. A maximum atrazine value of 0.78 μg/L was measured in the Save estuary that flows into the river Danube.

Moltmann et al. (2007) evaluated 21 pesticide-active ingredients for their relevance to surface water pollution and assigned high priority to *p,p'*-DDT and atrazine. Low priority was declared for the still authorized substance 2,4-D and the banned substances aldrin, β-HCH, dieldrin, endosulfan, endrin, γ-HCH, malathion, methoxychlor, parathion-methyl, mirex, *p,p*-DDE, and trifluralin.

6 Food Contamination

The European Commission recently published a report on pesticide residues in foods of herbal origin (EU 2008). The report is based on a systematic investigation performed in 2006 and summarizes the results of periodic monitoring of 25 EU countries, including Norway, Iceland, and Liechtenstein. Within the reporting period, a total of 65,810 samples (covering fruits, vegetables, field crops, and pre-treated products, including baby food) were analyzed. In total, 8,929,360 measurements of 54,747 samples (17,535 from Germany) were performed. The number of single-substance analyses varied among member countries and spanned 45–683 chemicals. Overall, 345 pesticide-active ingredients and their metabolites were detected. In 54% of all samples (38.1% of which were from Germany), no residues were detected. Of all positive findings, 42% (56.5% in Germany) were in the range of the maximum residue levels (MRLs) defined by the EU for each substance and product and 4.4% (5.35% in Germany) exceeded the MRL. For analyses performed in single-food categories, the following percentage of samples did not show detectable pesticide residues: 96% for baby food, 76% for pre-treated food, 73% for crops, and 51% for fruits and vegetables. Furthermore, it became evident that an exceedance of EU MRLs was more frequently observed for products originating from developing countries, compared to products originating from the EU (rate, 6.4/100 vs. 2.2/100). When comparing 10-year monitoring data (1996–2006), the percentage of foodstuff showing no detectable pesticide contamination continuously decreased, starting from 64% in 1999 to 51.5% in 2006. The percentage of samples exceeding the EU MRLs increased from 3.0 to 5.5%. In considering the significance of these trends, one must also remember that during the 10-year period, in which data were collected, analytical methodologies were enhanced and detection limits were lowered. Of all analyzed samples, 27.7% were contaminated by two or more pesticide-active ingredients or their metabolites.

Member countries were requested to compile a list of 10 active ingredients that are most frequently detected in their food samples, in order of decreasing frequencies. In Germany, the fruit and vegetable category was generally contaminated by chemicals according to the following pattern: maneb group > iprodione > procymidone (all thought to be endocrine active). Crop components most commonly contained substances of the maneb group > deltamethrin (thought to be endocrine active). A violation of EU MRLs was observed for substances of the maneb group (0.31% of all samples), dimethoate (0.27% of all samples) and procymidone (0.09% of all samples).

Market basket analyses were also conducted. The market basket contained eight fruit, vegetable, and other crop products (aubergines, bananas, cauliflower, grapefruits, orange juice, peas, bell pepper, and wheat). Analytical results showed that 56.9% of all samples had no measurable pesticide residues. In addition, 40.8% of the samples contained residues below the EU MRL. Pesticide contamination exceeding the MRL was observed for only 2.3% of the commodities. Within the scope of these analyses, residues of 55 pesticides were analyzed in food samples. Active ingredients were measured with decreasing frequency in grapefruit (68%), bananas (55%), bell peppers (42%), aubergines (33%), wheat (27%), peas (21%), cauliflower (20%), and orange juice (10%). Violations of MRL values were observed for aubergines, bell pepper, grapefruit, and pea samples. Approved and potentially endocrine-disrupting active ingredients were identified in food samples at the following relative frequencies: procymidone (16.6%), iprodione (15.6%), chlorpyrifos (15.0%), chemicals of the maneb group (13.3%), pyrimethanil (11.5%), and triadimenol (6.8%) in grapefruits. Aubergines contained predominantly procymidone (7.5% of all positive samples) and substances of the maneb group (6.8% of all samples). Bananas were mainly contaminated by chlorpyrifos (9.5% of all samples), peas by procymidone, bell peppers by procymidone and substances of the maneb group (14.0 and 9.2% of all samples, respectively), and similarly cauliflower by maneb group chemicals (29.5% of all positive samples).

Assessment of the potential chronic health risks associated with consuming contaminated foodstuffs was performed using the EFSA (European Food and Safety Authority) model. This model allows evaluators to consider country-specific eating habits. For 44 of 55 measured substances, the 90th percentile was below 0.01 mg/kg (general requirement for pesticide residues in food samples when specific limits are not provided by EU regulation). For these pesticides, a negligible uptake was expected. For four actual and potentially endocrine-active ingredients (chlorpyrifos, iprodion, maneb group, and procymidone), the 90th percentile level exceeded 0.01 mg/kg. These substances were checked to ascertain whether or not the approved ADI (acceptable daily intake) values were approached. In no case were the ADI values exceeded, because the substance exposure was lower than 0.9% of the ADI.

Acute risk assessment was performed for 34 of the 55 chemicals for which the Acute Reference Doses (ARfD) were defined by either the European Commission, the EFSA or the JMPR (Joint FAO/WHO Meeting on Pesticide Residues). Because only the maximum values were considered for calculating this risk assessment, results showed that the ingestion of a pooled food sample would have resulted in an ARfD exceedance for 15 of the 34 active ingredients. The following potential endocrine disruptors were among those 15 active ingredients: aldicarb, carbaryl, endosulfan, methomyl, parathion (banned chemicals) and dimethoate, λ-cyhalothrin, substances of the maneb group, and procymidone (approved chemicals). The number of samples exceeding the ARfD was comparatively low for those containing aldicarb, endosulfan, λ-cyhalothrin, parathion, dimethoate, and substances of the maneb group (1–5 samples) but was manifold higher for those contaminated with carbaryl, methomyl, and procymidone (14–20 samples). The most pronounced carbaryl ARfD violation was observed for grapefruit consumption, with values up to 464% (adults) and 956% (children) above the reference value. Distinct methomyl ARfD violations were recognized for bell pepper (up to 523 and 2,015% higher for adults and children, respectively) and grapefruit intake (adults up to 381% and children up to 786%). As a result, the EU withdrew the methomyl authorization in 2008 (re-registration in 2009). The highest ARfD exceedances for procymidone were noticed for grapefruit (up to 444% for adults and 917% for children). EU MRLs and more recent toxicological endpoints were checked by the Commission with regard to a prohibition of procymidone use (EU 2008; EFSA 2009). Actually, this substance is not approved under Annex I.

In Germany, the Federal Office of Consumer Protection and Food Safety (BVL) has carried out an independent Food Surveillance Programme since 1995. The program covers 72.5% of food samples of herbal origin (safflower and olive oil, rice, potatoes, spinach, onions, cucumber, green beans, carrots, red currant, peas, mandarins, apple juice, peppermint leaf tea, and rooibos tea) and 20% of food samples of animal origin (yoghurt, chicken meat, turkey meat, scalded sausages, salmon, cured trout filet, cured halibut, North Sea shrimp, and prawns) (BVL 2009). Ready-to-serve meals, candies (liquorice and chocolate), and baby food amounted to 7.5% of the sample size. The composition of the market basket utilized the Schroeter et al. (1999) model, in which German eating habits were considered. Of all samples in the market basket, 61% originated from Germany, 16% from EU member states, 13% from known, and 10% from unknown third countries. Samples were analyzed for residues of pesticides and other contaminants (biocides, veterinary drugs, heavy metals, etc.). In total, up to 52 pesticide-active ingredients, biocides, and metabolites were analyzed. In 2008, the monitoring program encompassed 5,093 samples.

Foods of animal origin predominantly contained persistent organic insecticide residues (e.g., p,p'-DDE, p,p'-DDD, p,p'-DDT, HCB, endosulfan sulfate, dieldrin, cis-/trans-nonachlor, cis-chlordane, oxychlordane, and toxaphene congener Parlar 26). Violations of MRLs were not detected. Samples characterized by having the most frequent positive findings were trout (74%), halibut (80%), and salmon (97%). Multiple pesticide residues (five active ingredients or more per sample) were particularly present in halibut (52% of all samples) and salmon (62% of all samples). Of the pesticide residues measured, 90% had residues below 0.05 mg active ingredients/kg.

Proportions equal to 27% of potatoes, spinach, onion, and apple juice retained pesticide residues. For safflower and olive oil samples, the quota of pesticide-contaminated samples was even lower and added up to 11%, although the BVL acknowledged that the samples were checked for comparatively few active ingredients (BVL 2009). Pesticide-active ingredients were more frequently found in rice, cucumbers, green beans, and carrots (59–70%). Of all rooibos tea samples 75% contained pesticide residues above the limit of detection. As in previous investigations, fruit revealed the highest incidence of positive pesticide findings (76–90% of the measurements exceeded the detection limit). Moreover, pears, red current, gooseberries, and mandarins presented the highest number of multiple pesticide residues (mean 3.3–3.9 active ingredients per sample).

Violation of MRL values was observed for 0.7–6.6% of spinach, onion, cucumber, green bean, red current, gooseberry, and mandarin samples. For rice, pears, and peppermint leaves, the exceedance quota amounted to more than 10% of the analyzed samples.

For single substances, the comparatively high MRL exploitation rates for carbendazim in rice, imazalil in mandarins, and amitraz in pears were noticeable (BVL 2009). For Turkish pears, a substantial exceedance of the amitraz (banned in EU member states) MRLs (and also ARfD values) was observed. Hence, these goods were withdrawn from sale (BVL 2009). Chemical-specific ARfD values were not affected for any other food sample of herbal origin.

No MRL exceedance was observed for olive and safflower oils, potatoes, carrots, apple juice, chocolate, or rooibos tea. MRL violations occurred in about 1.5% of home country samples, 1.8% of EU member state samples, and 17.9% of third country samples.

For 52 (9%) of all analyzed samples of German origin, the BVL assumed that pesticide-active ingredients were misused (BVL 2009). Such misused substances were mainly detected in peppermint leaves and pears. Residues that exceeded 0.01 mg/kg (lowest detection limit) were rated as indicating a non-approved application. However, BVL admits that this method did not allow them to differentiate between applications that were actually prohibited and applications of formerly approved persistent pesticides (brownfields), or seed and seedling treatments with banned foreign chemicals.

7 Conclusions

There are no generally accepted principles for what constitutes the critical avenues of pesticide loss from application or other sites. Such loss has many origins, including application technique, user expertise or experience, physicochemical properties of active ingredients applied, and local environmental conditions (precipitation quantity, soil quality, temperature, and average hours of sunshine per day). Therefore, quantifying pesticide loss via emission pathways varies considerably and depends on what monitoring data or mathematical computation models are used and the control variables that are applied (Table 2).

Although residue-free application is unrealistic, even very low residue concentrations may cause ecosystem damage as a result of multiple exposures or additive

Table 2 Pesticide release into the environment according to different routes of emission. Values originally provided in tons of emission per year (UBA 2000; Bach et al. 2005) have been converted on a percentage basis and refer to the total German pesticide consumption of approximately 30,000 t/a

Route of emission	UBA (2000)	Bach et al. (2005)	Carter (2000)
Spray drift (%)	~0.012	~0.00013	~0.3–3.5
Runoff (%)	~0.03	~0.05	~0.05
Volatilization	–	–	~20%
Drainage/leaching (%)	~0.005	~0.0006	~1
Point sources	~0.033%	~0.06%[a]	–

[a]Refers to emission into the aquatic environment only

effects, non-linear dose–response relationships, and susceptibility of organisms at sensitive life stages. Many pesticides that are suspected to have endocrine-disrupting properties have already been banned by the European Commission. Nevertheless, the realignment of the European Plant Health Legislation is not likely to solve the endocrine-disrupting properties that are associated with pesticide work, in part, because hormonal interferences may also result from mixture effects that are not addressed by the new EU legislation.

However, pesticide contamination has succeeded in attracting the attention of industry, agricultural enterprises, and authorities. Efforts have been made to reduce contamination by spray drift, e.g., by the development and implementation of advanced application techniques (low drift nozzles, air-assisted injector nozzles, etc.). Furthermore, the new EU directive 2009/128/EC (EU 2009b) binds all member states to ensure that the professional pesticide application equipment used is regularly inspected (5-year interval until 2020, thereafter 3 years). Finally, in the future, aerial spraying shall be allowed only in tightly controlled exceptional cases in all EU countries.

Directive 2009/128/EU addresses point source emissions through instructions that require training of professional users, including those who handle and store pesticides, clean equipment, or deal with remnant disposal. By December 2013, authorities are asked to establish certification systems to train professional pesticide users, distributors, and advisors (EU 2009b).

Preparation of this review chapter has suggested to the authors certain appropriate future action strategies that, if instituted, may help reduce pesticide residues in the environment. These include the following:

- Implementing a farmer advisory service independent of pesticide corporate interests;
- Fostering a broader embedding of water protection practices that will allow competence certification for agricultural pesticide users;
- Instigating an improved supra-regional information exchange on environmental pesticide contamination among (federal) regulatory authorities or other cooperating governmental or non-governmental groups;

- Developing a competitive pesticide classification system that will allow cultivators (farmers) and farm advisors to select the most eco-friendly pesticide for any specific authorized use;
- Assessing an eco-tax on pesticide products that will encourage use of minimal amounts of the proper product;
- Performing eco-audits of professional pesticide operators at regular intervals;
- Integrating a pesticide monitoring program for ground and surface waters on a nation/EU-wide basis;
- Addressing remobilization of previous pesticide contaminations via sediments and extending and harmonizing pesticide EQS values under WFD demands; and
- Utilizing data from existing monitoring programs that is submitted during the pesticide approval process.

8 Summary

The European Parliament recently approved a new EU regulation aimed at eliminating the use of pesticides that have unwanted endocrine-disrupting properties. The test criteria for these chemicals are slated to be finalized by 2013. For this reason, in this review, we have evaluated the metadata of lists and databanks that address pesticides with potentially endocrine-disrupting properties, and have checked which of the 250 active ingredients currently in use in Germany are affected. Azoles, dithiocarbamates/carbamates, and pyrethroids were most frequently rated as endocrine-active ingredients. In Germany, assessments have shown that total environmental pesticide emission is equivalent to approximately 0.1% of total pesticide use. Courtyard drainage and field runoff are regarded to constitute the most important sources of pesticide emission into the aquatic environment. In addition, in several investigations of drinking- and groundwater contamination, various pesticide-active ingredients and their metabolites were confirmed to be contaminants. Water suppliers recorded the following pesticides or their metabolites as being most frequently detected in drinking water: atrazine, desethylatrazine, diuron, simazine, isoproturon, and its dichlobenil metabolite 2,6-dichlorobenzamide. Surface water contamination results mainly from substances that are no longer approved by EU pesticide regulation. The most frequently detected pesticides in streaming waters that are still authorized were bentazone, diuron, glyphosate, isoproturon, MCPA, mecoprop, metamitron, pendimethalin, and tebuconazole.

Pesticide residues in comestible goods of herbal origin are periodically detected in all EU member countries. The European Commission recently published results showing that 54% of all monitoring samples were devoid of positive findings. Of samples showing detectable residues, 42% were below, and 4.4% exceeded the EU MRLs. Monitoring data over a 10-year period revealed that the percentage of food stuff without detectable pesticide residues has continuously decreased from 64 to 51.5%. In Germany, herbal samples mainly contained residues of maneb, iprodion, procymidone and deltamethrin. Notwithstanding these detections, chronic health risk evaluations indicated that there were no violations of ADI values. However, for carbaryl, methomyl, and procymidone, ARfDs were exceeded substantially for intake of grapefruit and bell peppers. As a result, the EU withdrew the methomyl authorization in 2008 and revised procymidone guideline values.

Acknowledgments This work was financially supported by the BMBF (Federal Ministry of Education and Research) as part of the *start₂* project (http://www.start-project.de/english_2.htm). The authors are grateful to the whole project team and the external experts for sharing their thoughts and the constructive discussions. Furthermore, we would like to extend our sincere thanks to the reviewers of this chapter for their helpful comments and support.

References

Altmayer B, Twertek M, Paetzhold M, Laronche JS (2003) Einträge von Pflanzenschutzmitteln in Gewässer – Situation im Weinbau und Gegenmaßnahmen. Gesunde Pflanzen 6: 161–168

ARGE Elbe – Working Group for the Water Quality Preservation of the Elbe (2008a) Gewässergütebericht der Elbe 2006. Wassergütestelle Hamburg, http://www.arge-elbe.de/wge/Download/Berichte/06Guetebericht.pdf. Accessed Nov 2009

ARGE Elbe – Working Group for the Water Quality Preservation of the Elbe (2008b) Gewässergütebericht der Elbe 2007. Wassergütestelle Hamburg, http://www.arge-elbe.de/wge/Download/Berichte/07Guetebericht.pdf. Accessed Nov 2009

Bach M (1999) Einträge aus Punktquellen und Gewässerfrachten. UBA Texte 85/99, Umweltbundesamt, Berlin, 70 pp

Bach M, Huber A, Frede HG (2001) Input pathways and river load of pesticides in Germany – a national scale modelling assessment. Water Sci Technol 43:261–268

Bach M, Röpke B, Frede HG (2005) Pesticides in rivers – assessment of source apportionment in the context of WFD. Eur Water Manage Online 2:1–14. http://www.ewaonline.de/journal/2005_02.pdf. Accessed June 2009

Bavarian Environment Agency (2008) Pflanzenschutzmittel in der Umwelt. Bayerisches Landesamt für Umwelt, Augsburg, 12 pp

BKH Consulting Engineers, TNO Nutrition and Food Research (2000) Towards the establishment of a priority list of substances for further evaluation of their role in endocrine disruption – preparation of a candidate list of substances as a basis for priority setting. Final Report for the EU, DG Environment. Delft, Zeist, The Netherlands, http://ec.europa.eu/environment/docum/pdf/bkh_main.pdf. Accessed Jan 2009

BMELV – Federal Ministry of Food, Agriculture and Consumer Protection (2009) Neue rechtliche Regelungen für Pflanzenschutzmittel Auf EU-Ebene. http://www.bmelv.de/cln_045/nn_751174/DE/04-Landwirtschaft/Pflanzenschutz/Aktuelles/Pflanzenschutzmittel.html__nnn=true. Accessed Jan 2009

BMG and UBA – Federal Ministry of Health and Environment Agency (2006) Bericht des Bundesministeriums für Gesundheit und des Umweltbundesamtes an die Verbraucherinnen und Verbraucher über die Qualität von Wasser für den menschlichen Gebrauch (Trinkwasser) in Deutschland. BMG, UBA, Bonn/Dessau, 38 pp

BMG and UBA – Federal Ministry of Health and Environment Agency (2008) Bericht des Bundesministeriums für Gesundheit und des Umweltbundesamtes an die Verbraucherinnen und Verbraucher über die Qualität von Wasser für den menschlichen Gebrauch (Trinkwasser) in Deutschland. BMG, UBA, Bonn/Dessau, 45 pp

BVL – Federal Office of Consumer Protection and Food Safety (2009) Berichte zur Lebensmittelsicherheit 2008. Lebensmittelmonitoring. Gemeinsamer Bericht des Bundes und der Länder. Bundesamt für Verbraucherschutz und Lebensmittelsicherheit, Berlin

BVL – Federal Office of Consumer Protection and Food Safety (2010) List of authorized plant protection products in Germany with information on terminated authorizations (Date: October 2010). Bundesamt für Verbraucherschutz und Lebensmittelsicherheit, Braunschweig. http://www.bvl.bund.de/cln_007/nn_492012/DE/04__Pflanzenschutzmittel/00__doks__downloads/psm__uebersichtsliste,templateId=raw,property=publicationFile.pdf/psm_uebersichtsliste.pdf

Carter A (2000) How pesticides get into water – and proposed reduction measures. Pest Outlook 11:149–156

DHI Water & Environment (2007) Study on enhancing the endocrine disrupter priority list with a focus on low production volume chemicals. Revised report to DG Environment. ENV.D.4/ETU/2005/0028r, http://ec.europa.eu/environment/endocrine/documents/final_report_2007.pdf. Accessed Jan 2009

Dubus IG, Hollis JM, Brown CD (2000) Pesticides in rainfall in Europe. Environ Pollut 110:331–344

EFSA (2009) MRLs of concern for the active substance procymidone, taking into account revised toxicological reference values. EFSA Sci Rep 227:1–26. http://www.efsa.europa.eu/en/scdocs/doc/227r.pdf. Accessed Sept 2010

EU (2006) Proposal for a directive of the European parliament and of the council on environmental quality standard in the field of water policy and amending Directive 2000/60/EC. COM(2006) 987 final. European Commission, Brussels

EU (2008) Monitoring of pesticide residues in products of plant origin in the European Union, Norway, Iceland and Liechtenstein 2006. Commission staff working document. SEC (2008) 2902 final. European Commission, Brussels

EU (2009a) Regulation (EC) No 1107/2009 of the European parliament and of the council of 21 October 2009 concerning the placing of plant protection products on the market and repealing council directives 79/117/EEC and 91/414/EEC. Off J E U 52:1–51

EU (2009b) Directive 2009/128/EC of the European parliament and the council of 21 October 2009 establishing a framework for community action to achieve the sustainable use of pesticides. Off J EU L 309:71–86

Eurostat (2007) The use of plant protection products in the European Union. Data 1992–2003. Office for official publications of the European Communities, Luxembourg

Fischer P, Hartmann H, Bach M, Burhenne J, Frede H-G, Spiteller M (1998) Gewässerbelastung durch Pflanzenschutzmittel in drei Einzugsgebieten. Gesunde Pflanzen 50:142–147

FOOTPRINT (2010) The FOOTPRINT pesticide properties database. Database collated by the University of Hertfordshire as part of the EU-funded FOOTPRINT project (FP6-SSP-022704), http://www.eu-footprint.org/ppdb.html. Accessed June 2010

Hessian State Parliament (2006) Official record of parliament 16/5841. Kleine Anfrage der Abgeordneten Ursula Hamman (Bündnis 90/Die Grünen) vom 14.07.2006 betreffend Belastung der Oberflächengewässer in Hessen durch Pestizide und Antwort des Ministers für Umwelt, ländlichen Raum und Verbraucherschutz, http://starweb.hessen.de/cache/DRS/16/1/05841.pdf. Accessed Nov 2009

HLUG – Hessian State Office for Environment and Geology (2010) Monitoring von Pflanzenschutzmittelwirkstoffen in hessischen Fließgewässern 2004 bis 2009, http://www.hlug.de/medien/wasser/wasser_psm/index.htm. Accessed Nov 2010

Huber A (1998) Belastung der Oberflächengewässer mit Pflanzenschutzmitteln in Deutschland – Modellierung der diffusen Einträge. Universität Gießen, Institut für Bodenkunde und Bodenerhaltung, 261 pp. ISBN-10: 3931789241

Huber A, Bach M, Frede HG (2000) Pollution of surface waters with pesticides in Germany: modelling non-point source inputs. Agric Ecosyst Environ 80:191–204

ICPDR – International Commission for the Protection of the Danube River (2002) Zusammenfassung des Endberichts Gemeinsame Donau-Untersuchung Mai 2002. ICPDR, Vienna

ICPR – International Commission for the Protection of the Rhine (2007) Stoffliste Rhein 2007. Report number 161d, http://www.iksr.org/uploads/media/Bericht_Nr._161d.pdf. Accessed Nov 2009

Kemikalieinspektionen (2008) Interpretation in Sweden of the impact of the "cut-off" criteria adopted in the common position of the council concerning the regulation of placing
plant protection products on the market (document 11119/08), http://www.kemi.se/upload/
Bekampningsmedel/Docs_eng/SE_positionpapper_annenII_sep08.pdf. Accessed May 2009

Kiefer J, Sturm S (2008) Pflanzenschutzmittel-Wirkstoffe und Metaboliten. Zusammenstellung
der häufigsten Funde in Oberflächen-, Grund- und Trinkwasser. Expert Contribution to:
Groundwater databank Water Supply, http://www.grundwasserdatenbank.de/bilder/pdf/TZW_
Sonderbeitrag_PflanzenschutzmittelWirkstoffeUndMetaboliten.pdf. Accessed Aug 2009

LAWA – Working Group of the Federal States on Water Issues (1998) 1. Bericht zur
Grundwasserbeschaffenheit – Pflanzenschutzmittel. Kulturbuchverlag, Berlin

LAWA – Working Group of the Federal States on Water Issues (2004) 2. Bericht zur
Grundwasserbeschaffenheit – Pflanzenschutzmittel. Kulturbuchverlag, Berlin

LUBW – Environment Agency Baden Wuertemberg (2004) Gütebericht 2002. Entwicklung der
Fließgewässerbeschaffenheit in Baden-Württemberg – chemisch –biologisch – morphologisch.
LfU Baden-Wuerttemberg, Karlsruhe

LUWG – Environment Agency Rhineland-Palatinate (2006) Organische Spurenstoffe in rheinland-
pfälzischen Fließgewässern 1985–2003. Nachweise, räumliche und zeitliche Schwerpunkte,
Qualitätszieleinhaltung. LUWG Rhineland-Palatinate, Oppenheim

McKinlay R, Plant JA, Bell JNB, Voulvoulis N (2008) Endocrine disrupting pesticides: implications for risk assessment. Environ Int 34:168–183

Moltmann JF, Liebig M, Knacker T, Keller M, Scheurer M, Ternes T (2007) Gewässerrelevanz
endokriner Stoffe und Arzneimittel. Neubewertung des Vorkommens, Erarbeitung eines
Monitoringkonzepts sowie Ausarbeitung von Maßnahmen zur Reduzierung des Eintrags in
Gewässer. Final Report FKZ 2005 24 205 by order of the Federal Environment Agency, Dessau

MUGV – Brandenburg State Office for the Environment (2007) Umweltdaten aus Brandenburg.
Bericht 2007. Environment Agency Brandenburg, Potsdam. http://www.mluv.brandenburg.de/
cms/media.php/lbm1.a.2320.de/umdat_07.pdf. Accessed Oct 2009

Müller K, Bach M, Hartmann H, Spiteller M, Frede HG (2002) Point and non-point source pesticide contamination in the Zwester Ohm Catchment (Germany). J Environ Qual 31:309–318

Neumann M, Schulz R, Schäfer K, Müller W, Mannheller W, Liess M (2002) The significance
of entry routes as point and non-point sources of pesticides in small streams. Water Res 36:
835–842

Neumeister L, Reuter W (2008) Die Schwarze Liste der Pestizide. Spritzmittel, die prioritär ersetzt
werden müssen – eine Handlungsanleitung für Industrie, Landwirtschaft, Lebensmittelhandel,
Politik und Behörden in Deutschland. Studie im Auftrag von Greenpeace e.V.; Greenpeace,
Hamburg

Pesticides Safety Directorate (2008) Revised assessment of the impact on crop protection in the UK
of the 'cut-off criteria' and substitution provisions in the proposed regulation of the European
parliament and of the council concerning the placing of plant protection products on the market. Pesticides Safety Directorate, York, UK http://www.pesticides.gov.uk/uploadedfiles/Web_
Assets/PSD/Revised_Impact_Report_1_Dec_2008(final).pdf. Accessed June 2010

RPS-BKH Consulting Engineers, DHI Water and Environment, Kiwa Water Research (2002)
Endocrine disrupters: study on gathering information on 435 substances with insufficient
data. Final Report for the EU, DG Environment, B4-3040/2001/325850/MAR/C2. Delft, The
Netherlands. http://ec.europa.eu/environment/endocrine/documents/bkh_report.pdf#page=1.
Accessed June 2010

Schroeter A, Sommerfelde G, Klein H, Hübner D (1999) Warenkorb für das Lebensmittel-
Monitoring in der Bundesrepublik Deutschland. Bundesgesundheitsblatt 1:77–83

Seel P, Knepper TP, Gabriel S, Weber A, Haberer K (1996) Kläranlagen als Haupteintragspfad
von Pflanzenschutzmitteln in ein Fließgewässer – Bilanzierung der Einträge. Vom Wasser 86:
247–262

Sturm S, Kiefer J (2007) Erhebung zur aktuellen Gewässerbelastung mit Pflanzenschutzmitteln.
Energie/Wasserpraxis 4:30–33

Sturm S, Kiefer J (2009) Befunde von Pflanzenschutzmitteln in Grundwässern Deutschlands. http://www.grundwasserdatenbank.de/PSM.htm. Accessed Aug 2009

Sturm S, Kiefer J, Eichhorn E (2007) Befunde von Pflanzenschutzmitteln in Grund- und Oberflächenwässern und deren Eintragspfade. Bedeutung für die Wasserwirtschaft und das Zulassungsverfahren. In: DVGW-Technologiezentrum Wasser (TZW) Karlsruhe (ed) Pflanzenschutzmittel in Böden, Grund- und Oberflächenwasser – Vorkommen, Abbau und Zulassung, vol 31. Publication of the Water Technology Center Karlsruhe, Karlsruhe, pp 185–311

UBA (2000) Daten zur Umwelt, 7th edn. Erich Schmidt Verlag, Berlin, 380 pp

Wagener H-A, Schuster M-E (2007) Pflanzenschutzmittel. In: Bayerisches Landesamt für Umwelt (ed) Chemikalien in der Umwelt – Medium Wasser. Expert Conference on 3rd May 2007, Augsburg, Conference Proceedings, pp 55–65. http://www.lfu.bayern.de/umweltwissen/doc/uw_btb_7_chemikalien_umwelt_medium_wasser.pdf. Accessed June 2010

Index

D.M. Whitacre (ed.), *Reviews of Environmental Contamination and Toxicology*,
Reviews of Environmental Contamination and Toxicology 213,
DOI 10.1007/978-1-4419-9860-6, © Springer Science+Business Media, LLC 2011